Disclaimer

The publisher of this book is by no way associated with the National Institute of Standards and Technology (NIST). The NIST did not publish this book. It was published by 50 page publications under the public domain license.

50 Page Publications.

Book Title: An Evaluation of Automated Latent Fingerprint Identification Technology (Phase II)

Book Author: Michael D. Indovina; Vladimir N. Dvornychenko; Elham Tabassi; George W. Quinn; Patrick J. Grother; Stephen Meagher; Michael D. Garris

Book Abstract: The National Institute of Standards and Technology (NIST), with the cooperation of eight technology providers, performed a test of accuracy for searching latent fingerprints when using automatically extracted features and matching (AFEM). This test is Phase II of the Evaluation of Latent Fingerprint Technology (ELFT) project. The test was open to both the commercial and academic community, and participants included vendors of Automated Fingerprint Identification Systems (AFIS). This report provides the design, process, assumptions, limitations, results, observations and conclusions of the test.

Citation: NIST Interagency/Internal Report (NISTIR) - 7577

Keyword: fingerprints; latent fingerprint; fingerprint matchers; automated matchers; lights-out; ELFT; Phase II; matcher performance

NISTIR 7577

ELFT Phase II - An Evaluation of Automated Latent Fingerprint Identification Technologies

M. Indovina
V. Dvornychenko
E. Tabassi
G. Quinn
P. Grother
S. Meagher
M. Garris

(This page intentionally left blank.)

NISTIR 7577

ELFT Phase II - An Evaluation of Automated Latent Fingerprint Identification Technologies

M. Indovina
V. Dvornychenko
E. Tabassi
G. Quinn
P. Grother
M. Garris
Information Access Division / Image Group
Information Technology Laboratory

S. Meagher
Dactyl ID, LLC

April 2, 2009

U.S. Department of Commerce
Gary Locke, Secretary

National Institute of Standards and Technology
Patrick D. Gallagher, Deputy Director

Acknowledgements

The authors would like to thank the Department of Homeland Security's Science and Technology Directorate and the Federal Bureau of Investigation's Criminal Justice Information Services Division for sponsoring this work.

Disclaimer

Specific hardware and software products identified in this report were used in order to perform the evaluations described in this document. In no case does identification of any commercial product, trade name, or vendor, imply recommendation or endorsement by the National Institute of Standards and Technology, nor does it imply that the products and equipment identified are necessarily the best available for the purpose.

Technology Providers

This table lists the technology providers who participated in this study. The letter keys listed down the first column are used throughout the report to identify results from specific algorithms. The authors wish to thank the technology providers for their voluntary participation and contribution.

Key	Technology Provider Name
K1	Motorola, Inc.
L1	Sonda Technologies, Ltd.
M1	NEC Corporation
N1	Peoplespot, Inc.
O1	SPEX Forensics, Inc.
P1	Cogent, Inc.
Q1	L1 Identity Solutions
R1	BioMG, Ltd.

Table 1: SDK letter keys and the corresponding technology provider

Executive Summary

Introduction:

The National Institute of Standards and Technology (NIST), with the cooperation of eight technology providers, performed a test of accuracy for searching latent fingerprints when using automatic feature extraction and matching (AFEM). This test is Phase II of the Evaluation of Latent Fingerprint Technology (ELFT) project. The test was open to both the commercial and academic community, and participants included vendors of Automated Fingerprint Identification Systems (AFIS). This report provides the design, process, caveats, results, observations and conclusions of the test.

The primary objective of the test is to determine whether significant latent print examiner time savings can be achieved while maintaining accuracy by not performing manual encoding of the latent fingerprint features. Doing so would permit a greater workload to be processed in the same amount of time and would potentially open up new opportunities for better exploitation of latent fingerprint services in various applications.

The eight technology providers each submitted a Software Development Kit (SDK) containing a latent fingerprint and ten-print minutiae extraction algorithm, and a 1-to-many match algorithm that returns a candidate list report. The specific fingerprint features extracted by the SDK were at the discretion of the technology provider and could be proprietary, and the feature template input to the SDK's matcher may include the original latent fingerprint image in its entirety. Technology providers were encouraged to submit research algorithms in this study. There was no requirement for the SDKs to be in operational use or commercially available. NIST performed a pre-test of the SDKs to ensure all functional capabilities were working. After validation of the SDKs, the technology providers were no longer involved in the testing. NIST performed the same test on all SDKs.

The test dataset contained 835 latent fingerprints, the associated ten-print fingerprint records containing the mates to the latent fingerprints, and two separate galleries of ten-print fingerprint records: one containing 5,000 records (50,000 fingerprints), and the second containing 10,000 records (100,000 fingerprints). The latent fingerprints were studied at two image resolutions: 1000 pixels per inch (ppi) (39.37 pixels per millimeter (ppmm)) images, and sub-sampled 1000 ppi producing 500 ppi (19.69 ppmm) images. In all tests, the ten-print galleries were 500 ppi[1]. The technology providers had no knowledge of, or access to, the fingerprint datasets prior to, during, or after the tests.

The SDKs were tested as black boxes. For each SDK, all ten-print fingerprint records and latent fingerprint images were processed by each SDK's automatic feature extraction algorithm. There was no human intervention during these processes. The automatically extracted features for the latent fingerprints were independently searched against the galleries of ten-print fingerprint features. A candidate list report was generated for each latent fingerprint search listing the top 50 candidates in ranked order by score, with the candidate having the highest score listed at rank 1.

In addition to assessing the overall performance of AFEM latent fingerprint technology, tests were designed to study specific factors expected to significantly impact performance. Insights into the effect of some of these factors may contribute to automated determination of latent fingerprint image quality. To this end, factors analyzed included the effect of gallery size, latent image resolution, supplementary region of interest, latent minutiae count, finger position, and finger pattern classification.

---

[1] Pixels per inch (ppi) is used throughout this report as this unit is commonly used across the biometric community, which is the audience to which this report has been written.

Caveats:
1. The 835 latent fingerprints represent identifications made using operational AFIS technology in actual case examinations. As a result, the latent fingerprints and their ten-print mates possess sufficient quality and quantity of information to result in identification, and therefore the results are representative of a category of higher quality latent fingerprints.
2. The characteristics of a latent print dataset that determine its difficulty level with respect to matching are largely determined by the source, selection, and preparation of the data. The results reported in this study may differ greatly from other latent datasets and operational fingerprint repositories.
3. The digital images of the latent fingerprints used for the test have undergone pre-search processing typical of AFIS operations. These include a combination of latent print analysis, selection criteria for AFIS search, scanning, orientation, image enhancement, classification, and finger designation.
4. The SDKs were not overly constrained by time in either extracting features or searching the galleries. It is possible for tighter time constraints to cause a decrease in performance. It is also possible to mitigate this concern by adding computing resources. However, the impact of time on performance was not tested.
5. The latent fingerprints were all directly captured at 1000 ppi. The creation of the 500 ppi images for the tests were produced by down-sampling the 1000 ppi images. The performance of matching with latents directly captured at 500 ppi was not tested in Phase II.

Results:

NIST performed analyses of the data and determined the performance and accuracy for each technology provider's SDK. A summary of identification rates based on candidate list position (rank) is reported in the following table. Note that each latent fingerprint search generated a list of fifty candidates, and it was generally observed that most identifications occurred within the top ten. Therefore, rank one and rank top ten results are reported.

SDK	Technology Provider	1000 ppi latents vs. 100K fgpts, Rank 1	1000 ppi latents vs. 100K fgpts, Rank 10	500 ppi latents vs. 50K fgpts, Rank 1	500 ppi latents vs. 50K fgpts, Rank 10
M1	NEC	97.2	98.8	96.4	97.2
P1	Cogent	87.8	89.2	88.0	89.9
O1	SPEX	80.0	85.6	80.0	87.1
K1	Motorola	79.3	83.2	79.6	84.0
Q1	L1 Identity Solutions	78.8	86.5	81.4	88.0
N1	Peoplespot	67.9	77.8	68.5	79.0
L1	Sonda	28.5	30.9	76.0	83.0
R1	BioMG	27.5	30.2	74.0	80.5

Table 2: Summary of Identification Rates (%)

Score-based measures can be used for two purposes: for candidate list reduction (eliminating low-probability candidates from candidate lists), and for automatic determinations of high-likelihood hits.

Candidate list reduction offers a tradeoff of accuracy for a reduction in human examiner workload: if a candidate list is reduced, the matcher will present shorter (or empty) candidate lists to the examiner, but some true mates will be excluded, lessening overall accuracy. This is illustrated by analyzing the results for the highest performing SDK shown in the Figure 1 (note that these results are based on probability of true mate[2] score values). At a false positive identification rate (FPIR) of 95% the false negative identification rate (FNIR) is 3%, whereas at FPIR of 47% the FNIR is 4%. Moving from the first operational point to the second cuts the examiner workload by up to half (FPIR from 95% to 47%), while missed identifications are increased by one third (FNIR 3% to 4%). It is a policy issue to determine if this is an acceptable trade-off.

---

[2] See observation 9.

Automatic determinations of high-likelihood hits can be used operationally to flag likely matches in low-priority cases that might otherwise never warrant examiner time, or to prioritize an examiner's workload based on the likelihood of match. In either case, automatic determinations of high-likelihood hits could be used for areas with an excessive backlog to maximize examiner efficiency. This is illustrated by analyzing the highest performing SDK shown in the figure below. At FNIR of 8% the FPIR is 1%. At this operating point, identifications are successfully made 92% of the time with only 1% of the examiner's comparisons including non-mates.

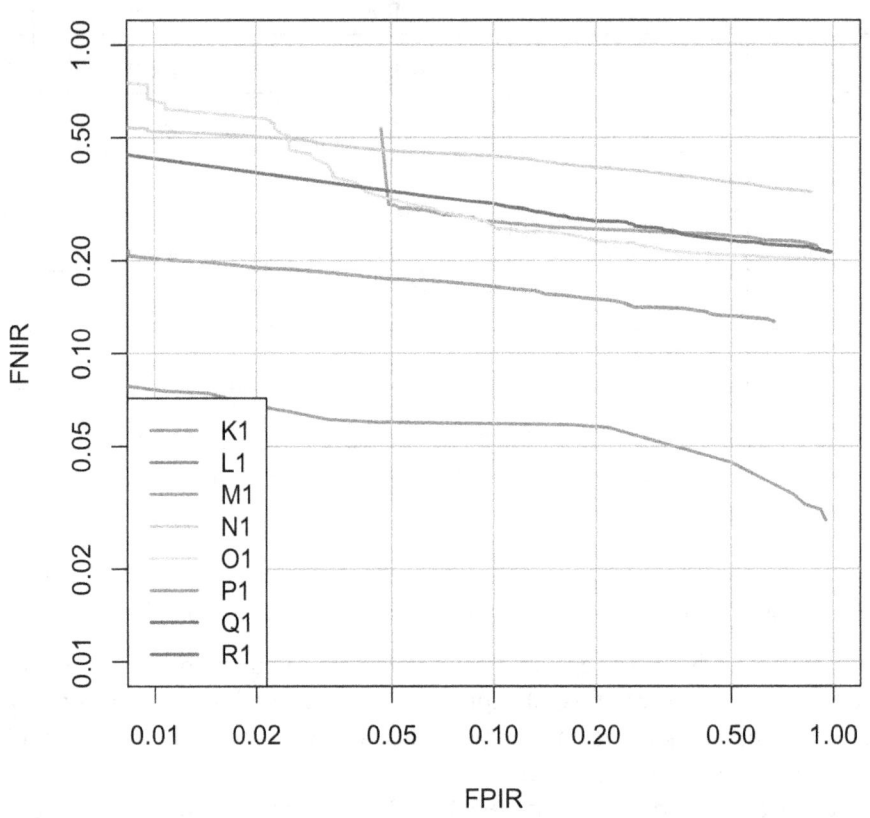

K1	Motorola	L1	Sonda	M1	NEC	N1	Peoplespot
O1	SPEX	P1	Cogent	Q1	L1 Identity Solutions	R1	BioMG

Figure 1: **Detection error trade**-off (DET) characteristics at rank 1

Observations:

1. The effect of scaling the gallery size was clearly observed as all SDKs demonstrated a drop in performance. The average decrease in rank 1 identification rate was 1% between searching a gallery of 50,000 fingerprints and a gallery of 100,000 fingerprints.
2. Five of the eight SDKs showed some benefit when searching latent images at 1000 ppi over 500 ppi (against a gallery of 500 ppi ten-prints), but the benefit was not shown to be statistically significant. An average improvement of 0.93% in rank 1 identification rate was observed. It was also observed that for every SDK, increasing resolution from 500 ppi to 1000 ppi caused some hits to be gained, but also some to be lost. The net outcome differs with each SDK.
3. The "region of interest" (ROI), produced mixed results. While for some SDKs the overall results improved when using ROI, for others they were worse. It was observed that those images with heavy excision (greater than 50% of the image cut out) tended to benefit the most. We conclude that ROI is inexpensive compared to manual markup, but so far has only been shown to be of limited use. More studies in this area are needed.
4. There is a strong correlation between the number of minutiae exhibited by a latent image and its search performance across all SDKs. Searches with higher numbers of minutiae tended to do better.
5. As with other biometric modalities, the quality of the image data strongly influences accuracy.
6. Looking at just the thumb, index, middle, and ring fingers on each hand, there is some evidence that latent search performance is highest with thumbs, next with index fingers, and lowest with ring fingers. However, results varied across the SDKs. Results on little fingers were not analyzed due to very small sample size representation in the Phase II dataset.
7. There is some evidence that latent search performance is affected by latent pattern class. Latent search performance was higher with whorls. The results for arches were bipolar; some SDKs performed best on arches; while other SDKs performed worst. Loops achieved medium performance. The undetermined category performed worst over all, which appears to be an indication of low latent image quality.
8. Fusing the latent search results (candidate lists) of multiple (cross-vendor) SDKs did improve the hit rate. The same is true for fusing the results of the same SDK for two or more of an individual's latent fingerprints. These methods provide a powerful mechanism for potential improvements of accuracy.
9. In addition to proprietary matcher scores, all SDKs reported a form of normalized scores, probability of true mate values in the range 0 to 100 indicating the SDK's estimated likelihood that a candidate is actually a mate. This has important implications for candidate list reduction, interoperability, and fusion. Results were mixed, with two SDKs demonstrating enhanced capability to reduce false matches.

Conclusions:

1. The results from ELFT Phase II demonstrate that a limited class of latent fingerprint case work can benefit from today's AFEM technology, thereby reducing some of the human workload during the AFIS latent fingerprint processes.

    Cautionary Note: Technology providers were encouraged to submit research algorithms in this study. There was no requirement for the SDKs to be in operational use or commercially available.

2. While the testing has demonstrated a level of performance beyond pre-test expectations, the limitations of the technology remain undefined and further testing is required.
3. Test results do not provide sufficient insight to determine with any specificity which latent prints in a case can benefit or should not be considered for AFEM. Latent fingerprint image quality measures are needed and should be tested.

Table of Contents

ACKNOWLEDGEMENTS 2
DISCLAIMER 2
TECHNOLOGY PROVIDERS 3
EXECUTIVE SUMMARY 4
TABLE OF CONTENTS 9
TERMS AND DEFINITIONS 10
1 INTRODUCTION 11
   1.1   ELFT AND AUTOMATED FEATURE EXTRACTION AND MATCHING 11
   1.2   PHASE I 11
   1.3   PHASE II 12
   1.4   WHAT PHASE II IS NOT 12
2 TEST IMPLEMENTATION 13
   2.1   EXPERIMENTAL DESIGN 14
   2.2   DATASET DESCRIPTION 16
      2.2.1   Latents 16
         2.2.1.1   Latent Pattern Classification 16
         2.2.1.2   Latent Finger Position 17
      2.2.2   Foreground 17
      2.2.3   Background 19
      2.2.4   Ten-Print Image Quality 19
      2.2.5   Mate Association and Validation 20
      2.2.6   ROI Markup 20
3 RESULTS AND ANALYSIS 22
   3.1   ACCURACY RESULTS OF PHASE II 22
      3.1.1   CMC 22
      3.1.2   DET 24
   3.2   COMPARISON WITH PHASE I RESULTS 30
   3.3   EFFECT OF GALLERY SIZE 31
   3.4   EFFECT OF RESOLUTION 33
   3.5   EFFECT OF REGION OF INTEREST (ROI) MARKUP 35
   3.6   EFFECT OF MINUTIA COUNT 38
   3.7   EFFECT OF FINGER POSITION 42
   3.8   EFFECT OF PATTERN CLASS 45
   3.9   TIMING RESULTS 48
4 PROBABILITY SCORES VS. RAW SCORES 53
   4.1   REPORTED PROBABILITY SCORES 53
   4.2   RE-COMPUTING DET USING PROBABILITY SCORES 55
5 FUSION 57
   5.1   MULTI-INSTANCE FUSION 57
   5.2   MULTI-ALGORITHM FUSION 62
6 REFERENCES 63
APPENDIX A – MODELLING EFFECT OF GALLERY SIZE 64
APPENDIX B – ELFT PHASE II PROTOCOL DESCRIPTION 70
APPENDIX C – ELFT PHASE II API SPECIFICATION 77
APPENDIX D - COMPLETE SET OF ACCURACY CHARACTERISTICS 88

Terms and Definitions

This table provides ELFT-specific definitions to various words and acronyms found in this report.

#	Term	Definition
1	AFEM	Automated Feature Extraction and Matching
2	API	Application Programming Interface
3	Background	A set of enrolled ten-prints not containing mate fingerprints
4	CMC	Cumulative Match Characteristic
5	DET	Detection Error Tradeoff characteristic
6	Exemplar	Fingerprint image acquired during an enrollment process and the mate of a latent fingerprint
7	FNIR	False Negative Identification Rate (also called miss rate or false non-match rate)
8	Foreground	A set of enrolled ten-prints containing mate fingerprints.
9	FPIR	False Positive Identification Rate (also called false-match rate)
10	Fusion	A method of combining biometric information to increase accuracy
11	Gallery	A set of enrolled ten-prints; synonymous with "database." An ELFT Gallery is composed of foreground and background ten-prints.
12	Hit/hit-rate	A "hit" results when the correct mate is placed on the candidate list; the "hit rate" is the fraction of times a hit occurs, assuming a mate is in the gallery.
13	Latent	A fingerprint image left on a surface touched by an individual
14	Matcher	Software functionality which produces one or more plausible candidates matching a search print
15	Mate	An enrolled fingerprint corresponding to a latent
16	NIST	National Institute of Standards and Technology
17	PPI	Pixels per inch (500 ppi corresponds to 197 pixels per centimeter)
18	ROC	Receiver Operator Characteristic
19	ROI	Region of Interest
20	Rolled print	A fingerprint image acquired by rolling a finger from side to side

Table 3: Glossary of ELFT Phase II related terms

1 Introduction

NIST has been investigating automated fingerprint matching since 1969, beginning with the pioneering work of Ray Moore. Focus on latent fingerprints is more recent, and was initiated in 2004 by a study comparing the performance of matching latent images against plain impressions versus rolled impressions [1]. In 2006, NIST inaugurated a more extensive project, called Evaluation of Latent Fingerprint Technology(ELFT) to investigate the performance of automated feature extraction and matching (AFEM) in the context of latent fingerprint identification [2].

ELFT is a study of latent fingerprint identification (one to many search) rather than verification (one to one match). Generally, the unknown fingerprint presented to an identification system may be any of the three types (rolled, plain, or latent), and the database against which it is searched may also be any of the three types, or even a mixture of types. In this study, the unknown is always a latent fingerprint, and the database consists entirely of rolled fingerprint impressions.

1.1 ELFT and Automated Feature Extraction and Matching

It is important to distinguish AFEM-based latent fingerprint identification from the general concept of lights-out identification. Lights-out identification refers to a system requiring minimal or zero human assistance in which an image is presented as input, and the output consists of a short candidate list. For ten-print search applications, this list may be: 1) empty, 2) contain a single candidate, or rarely 3) have more than one candidate. Event (3) will occur only in cases when the matcher produces more than one candidate with a significant computed probability of being a true mate. Lights-out matchers are currently in operation for rolled fingerprint search systems, and are emerging for plain impressions. Latent fingerprints are much more difficult, and no lights-out matchers are currently in operational service. Furthermore, the ELFT07 Concept of Operations (CONOPS) asserts that a fully lights-out latent fingerprint matching capability represents too large of a single step from current practices [2].

Accordingly, the initial focus of ELFT is on AFEM-based latent fingerprint identification systems in which manual feature extraction by an examiner is eliminated (i.e. the feature extraction and search operations are fully automated), but the candidate lists output by these systems may be of non-trivial size, and require varying degrees of human inspection. This automates the traditional human feature selection on the latent image – often referred to as "front end functions" – but does not fully address the "backend" functions, including reduction of the output candidate list. Human feature extraction by a latent fingerprint examiner is a time-consuming task. It is common for the examiner to spend twenty minutes or more on this step. It is therefore highly desirable to automate feature extraction to the extent possible, as well as automate any other time-consuming steps.

In current latent matching practice the candidate lists tend to be of fixed length, typically 10 to 20 candidates long. A fixed number of candidates is then always produced, even though the vast majority of these are non-matching to the search latent. (Although they do represent the best matches encountered, these are not close matches in any meaningful sense and cursory inspection often reveals that the candidate cannot possibly match the search print.) One goal of ELFT is to suppress these non-matching candidates, resulting in a much shorter candidate list of variable length. We refer to this goal as candidate list reduction.

1.2 Phase I

ELFT Phase I was designed as a "proof of concept" test for evaluating state-of-the-art AFEM-based one-to-many (1:N) latent fingerprint identification systems. A secondary goal of Phase I was to determine the test methods and metrics necessary to evaluate the technology. Participation was open to all, and testing and reporting was done anonymously to encourage participation and minimize risk to participants. Software was submitted to NIST for testing in the form of Software Development Kits (SDKs) conformant to an Application

Programming Interface (API) created by NIST. Ten technology providers participated in Phase I, submitting a total of 16 SDKs for testing. These SDKs were installed and executed on NIST hardware, by NIST personnel. The dataset used for Phase I was a mix of operational and non-operational images. A total of 100 latent images were searched against a gallery of 10,000 rolled fingerprints (1000 ten-prints). The aggregate results of Phase I have been reported publically without mention of the technology provider names [4], and detailed reports have been provided directly to the individual participants. Phase I demonstrated the feasibility of the technology and test methods, and provided valuable insights into how future phases should be conducted.

1.3 Phase II

Whereas ELFT Phase I was intended to assess the feasibility of AFEM-based latent fingerprint identification systems, ELFT Phase II was designed to assess the performance of state-of-the-art AFEM technology and evaluate its viability for operational use. ELFT Phase II builds on the work in Phase I by using 100% operational images, as well as larger and more diverse datasets to provide better performance estimates. A primary objective of the test was to determine whether significant latent print examiner time savings can be achieved by applying AFEM technologies while maintaining accuracy. Doing so would permit a greater workload to be processed in the same amount of time and would potentially open up new opportunities for better exploiting latent fingerprint services in various applications.

A further objective of Phase II was to study specific factors that are expected to significantly impact the performance of AFEM latent fingerprint technologies. It is anticipated that insights into the effect of these factors may contribute to automated determination of latent fingerprint image quality, which is a key component to future AFEM-based systems. Phase II analyzed the effect of the following factors:

- Gallery size
- Latent image resolution
- Supplementary region of interest (ROI)
- Latent minutiae count
- Finger position
- Finger pattern classification

In all, eight technology providers chose to participate in ELFT Phase II, each contributing an SDK for testing. Section 2 provides an overview of the Phase II test implementation, experimental design, and the dataset used. Section 3 reports accuracy results and analyses on various factors. Section 4 discusses the results of using probability scores in place of raw matcher scores. Section 5 presents a study on candidate list fusion. The appendices cover topics including a simple model for predicting the effect of increasing gallery size along with the ELFT Phase II protocol and application program interface (API). A complete set of Detection Error Trade-off (DET) curves created for this study are included in Appendix D.

1.4 What Phase II is Not

The following are specifically not within the scope of ELFT Phase II:

- Evaluation of human examiner assisted latent fingerprint based identification
- Evaluation of standardized fingerprint feature encodings and standard templates
- Closed set ("closed universe") identification
- Verification or one-to-one (1:1) matcher performance
- Performance when matching ten-print records against a repository of latent fingerprints (also called "reverse latent" searches)
- Performance when matching latent fingerprints to latent fingerprints (latent-to-latent)
- Evaluation of latent fingerprint collection/processing methods
- Estimates of algorithm speed when implemented in operational systems
- Template update or adaptive search algorithms

2 Test Implementation

The testing model used by ELFT is similar to that used by NIST for Minutiae Interoperability Exchange (MINEX) Test 2004 [5] and Proprietary Fingerprint Template Test [6]. Binary software modules only (no source code), referred to as SDKs, are solicited from participants. These are intended to be executed by NIST on local computer hardware, and they must adhere to an Application Programming Interface (API) specified by NIST.

The SDK testing model is different from other fingerprint evaluations such as Fingerprint Vendor Technology Evaluation (FpVTE) [7] and Fingerprint Verification Competition (FVC) [8] in that it provides greater flexibility and control over the execution of software during the test. Testing of the SDKs on NIST hardware by NIST personnel ensures that the test images themselves are never disclosed outside of NIST. This has important implications for privacy as well as for the use of these images in future testing. The principal disadvantage of this approach is that it limits the feedback NIST can provide to participants regarding image-specific behavior of their software.

Each technology provider submitted an SDK containing a latent fingerprint and ten-print minutiae extraction algorithm, and a 1-to-many match algorithm that returns a candidate list report. The specific fingerprint features extracted by the SDK were at the discretion of the technology provider and could be proprietary, and the feature template input to the SDK's matcher may include the original latent fingerprint image in its entirety. Technology providers were encouraged to submit research algorithms in this study. There was no requirement for the SDKs to be in operational use or commercially available. NIST performed a pre-test of the SDKs to ensure all functional capabilities were working. After validation of the SDKs, the technology providers were no longer involved in the testing. NIST performed the same test on all SDKs.

The test dataset contained 835 latent fingerprints, the associated ten-print fingerprint records containing the mates to the latent fingerprints, and two separate galleries of ten-print fingerprint records: one containing 5,000 records (50,000 fingerprints), and the second containing 10,000 records (100,000 fingerprints). The technology providers had no knowledge of, or access to, the fingerprint datasets prior to, during, or after the tests. Latent prints were searched at both 500 pixels per inch (ppi) (19.69 pixels per millimeter (ppmm)) and 1000 ppi (39.37 ppmm) resolutions, as well as with and without supplementary ROI markup. In all tests, the ten-print galleries were 500 ppi.

The SDKs were tested as black boxes. For each SDK, all ten-print fingerprint records and latent fingerprint images were processed by each SDK's automatic feature extraction algorithm. There was no human intervention during these processes. The automatically extracted features for the latent fingerprints were independently searched against both galleries of ten-print fingerprint features. A candidate list report was generated for each latent fingerprint search listing the top 50 candidates in ranked order by score, with the candidate having the highest score list at rank 1.

The computer hardware used to run the SDKs were of the following configuration:

- Processor:
    - Dual 2.8 GHz/1MB Cache, Xeon (dual-core)
    - 800 MHz Front Side Bus for PE 1855
- Memory:
    - 2GB DDR2 400 MHz (2x1GB) single ranked DIMMS
- Secondary storage:
    - DUAL 73GB 10K RPM, Ultra 320 80 pin SCSI Hard drives (hot plug)

Note that the above hardware may not be representative of operational systems. Therefore, all timing requirements and all timing results may not be directly comparable to operational scenarios, which may involve different hardware and software.

In the remainder of this section, section 2.1 presents the overall experimental design and section 2.2 describes the dataset of latents, ten-print exemplars (foreground), and ten-print background used.

2.1 Experimental Design

NIST ran each of the eight Phase II SDKs through a series of four major latent search tests. Each test involved a different configuration of latents and ten-prints designed to measure overall performance, as well as analyze the effects of gallery size (5,000 vs. 10,000 ten-print records), image resolution (500 ppi vs.1000 ppi), and supplementary ROI.

There were three different sets of latents used in the study; each consisted of impressions from the same 835 distinct fingers. These are listed in the table below. The first set, L1, consisted of images at 500 ppi resolution. The second, L2, consisted of images at 1000 ppi resolution. The third set, L3, consisted of the same 1000 ppi images in L2, but with ROI included in the search. All latent images were originally scanned at 1000 ppi resolution. The corresponding 500 ppi latents in L1 were produced by sub-sampling the 1000 ppi images in L2. None of the latent images were ever lossy-compressed. More details on the characteristics of the latent fingerprint images used in this study are provided in section 2.2.1.

Latent test set	Total latents	Latent image resolution (ppi)	ROI included?
L1	835	500	No
L2	835	1000	No
L3	835	1000	Yes

Table 4: Phase II latent test sets

There were four different configurations of galleries used in the study; the four are paired into two groups. These are listed in the table below. The first pair, G1A & G1B, contains a 10K random sample of ten-print records; the second pair, G2A & G2B, contains a 5K proper subset of those records in G1A & G1B. In both pairings, the 'A' set includes the mated ten-print records, called the foreground, while the 'B' set excludes the mated ten-print records, called the background. Another way to say this is the 'A' set is a seeded gallery comprised of background ten-print records in addition to the latent set's mated ten-print records. The 'B' set is unseeded, and contains none of the latent set's mated ten-print records. The seeded galleries have been constructed with the foreground ten-print records randomly distributed throughout. For each latent in a test set, there is one, and only one, mate in the corresponding seeded gallery. By searching a seeded gallery, genuine (true) match scores can be accumulated contributing to the calculation of the False Negative Identification Rate (FNIR). By searching an unseeded gallery, imposter (non-match) scores can be accumulated contributing the calculation of the False Positive Identification Rate (FPIR).

It should be noted that all ten-print records used in this study were scanned at 500 ppi, and they were lossy-compressed using Wavelet Scalar Quantization (WSQ). More details on the characteristics of the ten-print records used in this study are provided in sections 2.2.2 and 2.2.3.

Gallery test set	Total ten-print records	Mate ten-print records (foreground)	Non-mate ten-print records (background)	Ten-print image resolution (ppi)
G1A	10000	608	9392	500
G1B	10000	0	10000	500
G2A	5000	608	4392	500
G2B	5000	0	5000	500

Table 5: Phase II gallery test sets

The latent and gallery test sets, described above, were combined into the four major latent search tests listed in the table below. Notice that not all possible combinations of latent to gallery configurations were executed in Phase II. This was due to the large computation time it takes to execute one SDK on just one search configuration. It was determined that there was only time to execute the four tests below by each of the eight SDKs, and that the selection of these four was sufficient to study the effects of gallery size, image resolution, and supplementary ROI. Results in this report identify which of these four tests they were derived from, and analyses identify which of these four tests are studied and/or compared.

Latent search test	Search configuration	
	Latent test set	Gallery test set
500 ppi Latent vs. 5K Gallery	L1	G2A
	L1	G2B
1000 ppi Latent vs. 10K Gallery	L2	G1A
	L2	G1B
1000 ppi Latent vs. 5K Gallery	L2	G2A
	L2	G2B
1000 ppi Latent +ROI vs. 5K Gallery	L3	G2A
	L3	G2B

Table 6: Phase II major latent search tests

2.2 Dataset Description

2.2.1 Latents

The images NIST used for this evaluation were obtained from an unnamed U.S. Government operational source. The latents were prepared during the course of feature search transactions conducted by an examiner using the FBI's Integrated Automated Fingerprint Identification System (IAFIS). For each latent, a single examiner was involved in feature selection, searching, and subsequent pairing with a subject's ten-print mate. The complete set of latent images, minutiae features, images of ten-print mates, and original IAFIS candidate lists were provided to NIST, along with associated metadata.

The dataset contains latents of type "latent photo," and most are likely to have originated from paper medium after first being developed using ninhydrin (the precise methods of developing the print are not documented). The images themselves are assumed to be scans of original photographic prints, which make them 2nd generation images. Most all images were modified by the examiner conducting the original case work to enhance usability, which typically included cropping and/or rotation of the original scanned image. In approximately one third of the cases, a kind of "contrast enhancement" was applied which sometimes resulted in the images having a "binarized" appearance.

All latent images in the dataset were originally scanned at 1000 ppi resolution. The width of these images ranges from 216 to 1510 pixels, and the height ranges from 276 to 1667 pixels. The mean width is 586 pixels, and the mean height is 696 pixels.

2.2.1.1 Latent Pattern Classification

Table 7 lists the percentages of latent prints in the dataset belonging to three major ridge flow pattern classifications as reported to NIST: loops, whorls and arches. All pattern classifications listed here were determined by the examiners when conducting the original case work. (Additionally, a fingerprint examiner at NIST verified these original classifications.) A fourth category of "undetermined" is included for those latent prints in which there was no discernable pattern class. The table compares these percentages to the approximate pattern class percentages for all fingerprints in the FBI's Criminal Master File (CMF). Based on this, the ELFT dataset appears somewhat overrepresented in whorls and underrepresented in loops.

	ELFT Phase II Dataset (%)	FBI CMF (%)
Loops	46.8	65
Whorls	41.7	30
Arches	3.6	5
Undetermined	7.9	N/A

Table 7: Distribution of latent pattern classes for ELFT Phase II dataset vs. FBI CMF

2.2.1.2 Latent Finger Position

Table 8 lists the percentages of latent prints in the dataset by finger position as determined by the examiner during the original case work. Finger position of each latent is based on the finger position of the corresponding ten-print mate. As shown, the dataset contains predominately thumbs (47.2%) and index fingers (27.3%). It should be noted that the prevalence of whorls shown in Table 7 is most likely due to the dataset containing a large percentage of thumbs, as thumbs are more likely to have whorls.

Finger position	Right thumb	Right index	Right middle	Right ring	Right little	Left thumb	Left index	Left middle	Left ring	Left little
% of total	29.6	15.2	7.7	2.4	0.7	17.6	12.1	7.7	5.7	1.3

Table 8: Distribution of latent fingerprint positions in ELFT Phase II dataset

2.2.2 Foreground

For each latent image in the dataset, there is a corresponding ten-print mate image. The collection of ten-prints containing these mates comprises the foreground. While 588 unique subjects are associated with the latents in the dataset, the number of subjects associated with the foreground ten-prints is 608 – this is an artifact of a number of latents having been excluded from the study after gallery creation. (See section 2.2.5 for more detail.)

Figure 2 shows how the latent prints are distributed by subject. The majority of subjects (75%) have a single associated latent in the dataset; however, a significant percentage (25%) have two or more. The top chart in the figure shows the total number of latent images per subject in the dataset, whereas the bottom chart counts only distinct physical fingers present per subject (e.g., a subject having three total fingers where two are right index and one is right thumb is counted as having only two "distinct" fingers). This is of relevance to section 5, which examines the potential utility for "fusion" of search results when two or more fingers are available per subject.

The foreground consists of approximately 18% live-scanned and 82% non-live scanned (ink) impressions. The foreground images were all scanned at 500 ppi, and were WSQ compressed (15:1). The image width of the foreground images ranges from 304 to 837 pixels, and the height ranges from 432 to 800 pixels. The mean width is 798 pixels, and the mean height is 752 pixels

2.2.3 Background

The dataset consists of a larger number of non-mate ten-prints called the background. To ensure dataset diversity, the background was selected to be composed of ten-print images from four different government operational datasets: TXDPS, LACNTY, AZDPS, and INSBEN. Ten-prints were randomly selected from these datasets in order to achieve a 50/50 mix of ink and live-scan impression types along with equal proportionality from each source. Table 9 below shows the composition by data source of each gallery defined in Table 5. The background images were all scanned at 500 ppi, and were WSQ compressed (15:1). The image width of the background images ranges from 384 to 812 pixels, and the height ranges from 544 to 801 pixels. The mean width is 800 pixels, and the mean height is 742 pixels.

Gallery Name	Non-mate ten-print records (background)	TXDPS (100% ink)	LACNTY (50% live-scan and 50% ink)	AZDPS (100% live-scan)	INSBEN (50% live-scan and 50% ink)
G1A	9392	2348	2348	2348	2348
G1B	10000	2500	2500	2500	2500
G2A	4392	1098	1098	1098	1098
G2B	5000	1250	1250	1250	1250

Table 9: Phase II background sources

2.2.4 Ten-Print Image Quality

Table 10 summarizes the quality of the ten-prints by source comprising the ELFT Phase II dataset. The NIST Fingerprint Image Quality (NFIQ) [9] is being used for the comparison with NFIQ=1 of highest image quality and NFIQ=5 of lowest image quality. (Note that NFIQ was primarily designed for use with plain fingerprints, but is being applied to rolled prints in this case.) The Summary NFIQ listed in the right column of the table is based on equation 1 of NIST IR 7422 [10] and is on the scale of 0 to 100, with 100 being of highest image quality. As can be seen from Table 10, the resulting foreground ten-prints are generally of higher quality than the background ten-prints.

Dataset name	NFIQ=1 (%)	NFIQ=2 (%)	NFIQ=3 (%)	NFIQ=4 (%)	NFIQ=5 (%)	Summary NFIQ
TXDPS	36.5	9.2	38.7	6.4	9.2	83.6
LACNTY	33.4	17.8	29.5	5.0	14.2	79.9
AZDPS	38.5	17.3	24.2	9.4	10.6	82.4
INSBEN	31.0	15.2	29.9	11.8	12.2	79.3
ELFT foreground	44.1	15.5	27.9	4.7	7.8	86.7
ELFT background	34.9	14.9	30.6	8.1	11.5	81.3

Table 10: Summary of ten-print quality by source

2.2.5 Mate Association and Validation

Accurately determining the correspondence between latent fingerprints and their associated ten-print mates is essential for the purpose of measuring latent search error rates. As described earlier, the latent images used for this study are from real case work, and the metadata provided was used to establish the associations to the corresponding ten-print mates. Specifically, all latents in this study were originally submitted to the FBI IAFIS system in order to produce suitable candidate lists for subsequent (1:1) comparisons by US government latent fingerprint examiners. In each case, a ten-print record was eventually determined by examiners to contain a mate corresponding to the search latent.

Use of latent data obtained as a result of a successful AFIS search introduces "AFIS bias." AFIS bias results when the data has been selected using an AFIS. Latents that may have a mate in the database, but which are not matched because of poor exemplar image quality, relative distortion differences, or insufficient overlap, are excluded. This can result in a significant overstatement of the system performance capability as compared to the expected performance level that is likely to be achieved in an operational environment. The quality of the latent fingerprints used in Phase II was reviewed by a latent fingerprint examiner, and it was determined that the overall quality was higher than typical operational latent case work. This is supported by the observation that the average minutiae count for the Phase II dataset is 23, which is higher than the average minutiae count observed in typical latent case work (reported to be 17.[3])

In order to validate the association of identity between latents and their alleged mates, a limited number of cases were reviewed by a latent examiner at NIST. Included for review were cases in which all SDKs missed reporting the alleged mate on the candidate list, cases in which the highest performing SDK missed reporting the alleged mate on the candidate list, cases of high scoring alleged non-mates, and cases of low scoring alleged mates. Of all cases reviewed, six were removed as confirmed non-matches. Six cases were determined inconclusive, and of these, three were missed by all SDKs and therefore removed. For each of the three inconclusive cases retained in the dataset, the associated mate provided from the original case work was assumed to be true, and SDKs were scored accordingly. This validation process promoted the accuracy of testing while it minimized any bias in removing particularly difficult latents.

2.2.6 ROI Markup

The ELFT Phase II CONOPS proposed a type of human assistance to the AFEM SDKs requiring minimal effort on the part of the human expert. The basic concept is to have the examiner designate an ROI on the latent image. The ROI would include the area of the primary fingerprint in the image, while excluding areas of non-fingerprint data, extraneous fingerprints, and / or highly smudged parts of the fingerprint. In principle, this should increase the signal-to-noise ratio, resulting in better performance. Experience has shown that selecting an ROI is much less labor intensive than extracting minutiae manually, and can typically be done by an expert in well under a minute, compared to possibly 20 minutes for a full examiner-markup.

Figure 3 shows the steps in creating an ROI. In the first image (a) an ROI is outlined on the latent fingerprint image. The next figure (b) shows the resulting ROI mask by itself. This is followed by a figure (c), which shows the effect of applying the ROI mask to the image. These three images constitute a simplified example, for illustrative purposes. The actual Phase II masks had more complicated boundaries, as shown in (d) for a different latent image case.

The ROI is intended to preserve the "good" part of the fingerprint image, while excising regions considered possibly "detrimental." When creating the ROI masks, NIST found that there were many cases where excising any part of the image would be an arbitrary decision, and therefore unjustified. In these cases the ROI mask was selected to cover the entire image, and in essence the entire image was retained. Of the 835 latent cases in the dataset, 89 included an ROI of the entire latent image.

---

[3] Minutiae count of 17 for typical case work cited by NIST latent expert.

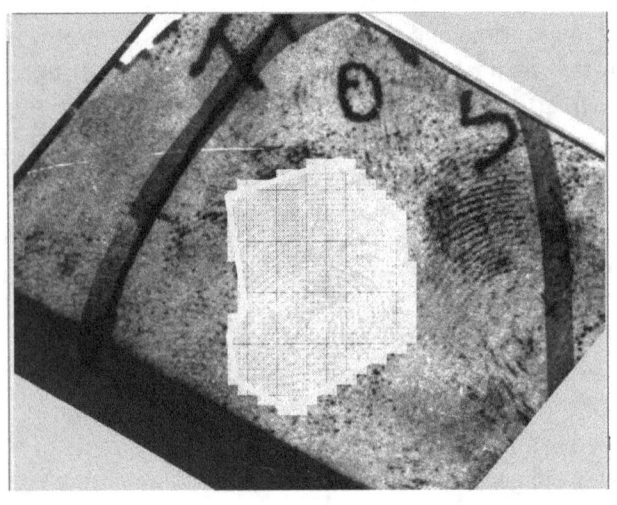

(a) Original image with region-of-interest marked

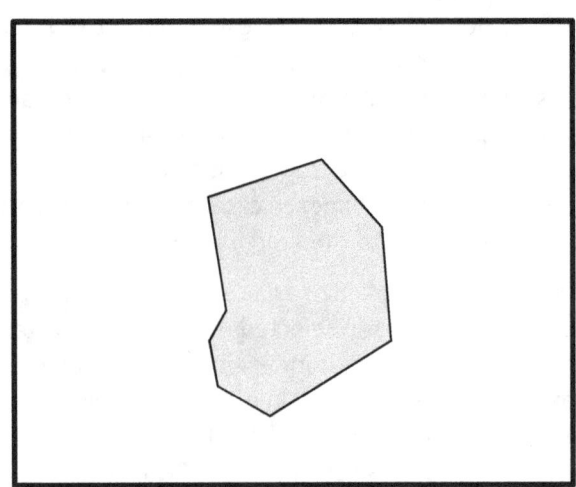

(b) Image mask created from region of interest

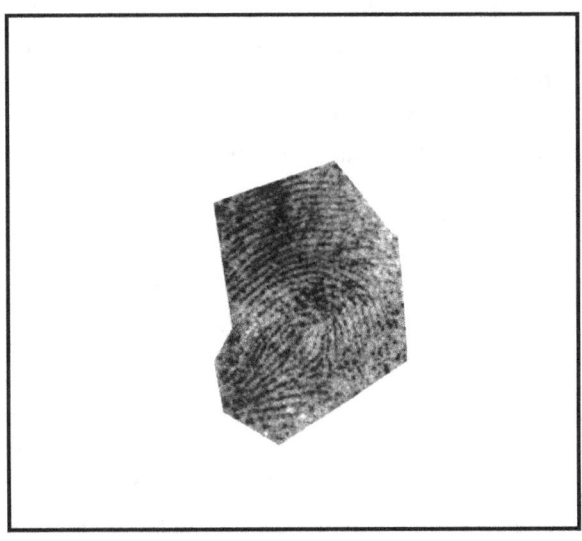

(c) Sub-mage created by applying mask (b) to original image (a)

(d) Specimen of actual Phase II mask (unrelated to prior image)

Figure 3: Example of Region-of-Interest (ROI) markup

3 Results and Analysis

3.1 Accuracy Results of Phase II

Overall accuracy results are presented in this section first using rank-based metrics via Cumulative Match Characteristic (CMC) curves, and then using score-based metrics via Detection Error Trade-off (DET) curves. It should be noticed that ranked-based statistics are not necessarily stable as database size grows.

3.1.1 CMC

A CMC curve shows how many latent images are correctly identified at rank 1, rank 2, etc. A CMC is a plot of identification rate (or hit rate) vs. recognition rank. Identification rate at rank k is the proportion of the latent images correctly identified at rank K or lower. A latent image has rank k if its mate is the kth largest comparison score on the candidate list. Recognition rank ranges from 1 to 50, as 50 was the (maximum) candidate list size specified in the API.

Figure 4 shows CMC plots of the eight SDKs for different gallery sizes (5K and 10K) and latent images at different resolution (500 ppi and 1000 ppi). SDK M1 outperforms other SDKs, with P1 at a distant second place. This is confirmed by Figure 5, which shows the identification rate at rank 1 (i.e. percentage of hits in first position) for the eight SDKs along with the 95% confidence intervals for latents at 500 ppi and gallery of 5K. The non-overlapping confidence intervals of M1 and P1 suggest significant difference between their performance and other SDKs. Figure 4 also suggests that there is not much gain in identification rate after rank 10. Also, L1 and R1 exhibit curiously lower performance on 1000 ppi than on 500 ppi latents. There are small changes observed between the CMC plots when comparing effect of gallery size, latent image resolution (ignoring L1 & R1), and supplementary ROI. These factors will be analyzed more closely in later sections.

Key Observations:

- SDK M1 outperforms other SDKs (rank 1 identification rate of 97.2% for 1000 ppi gallery 10K) ; the second best performer is P1.
- There is not much gain in identification rate after rank 10.

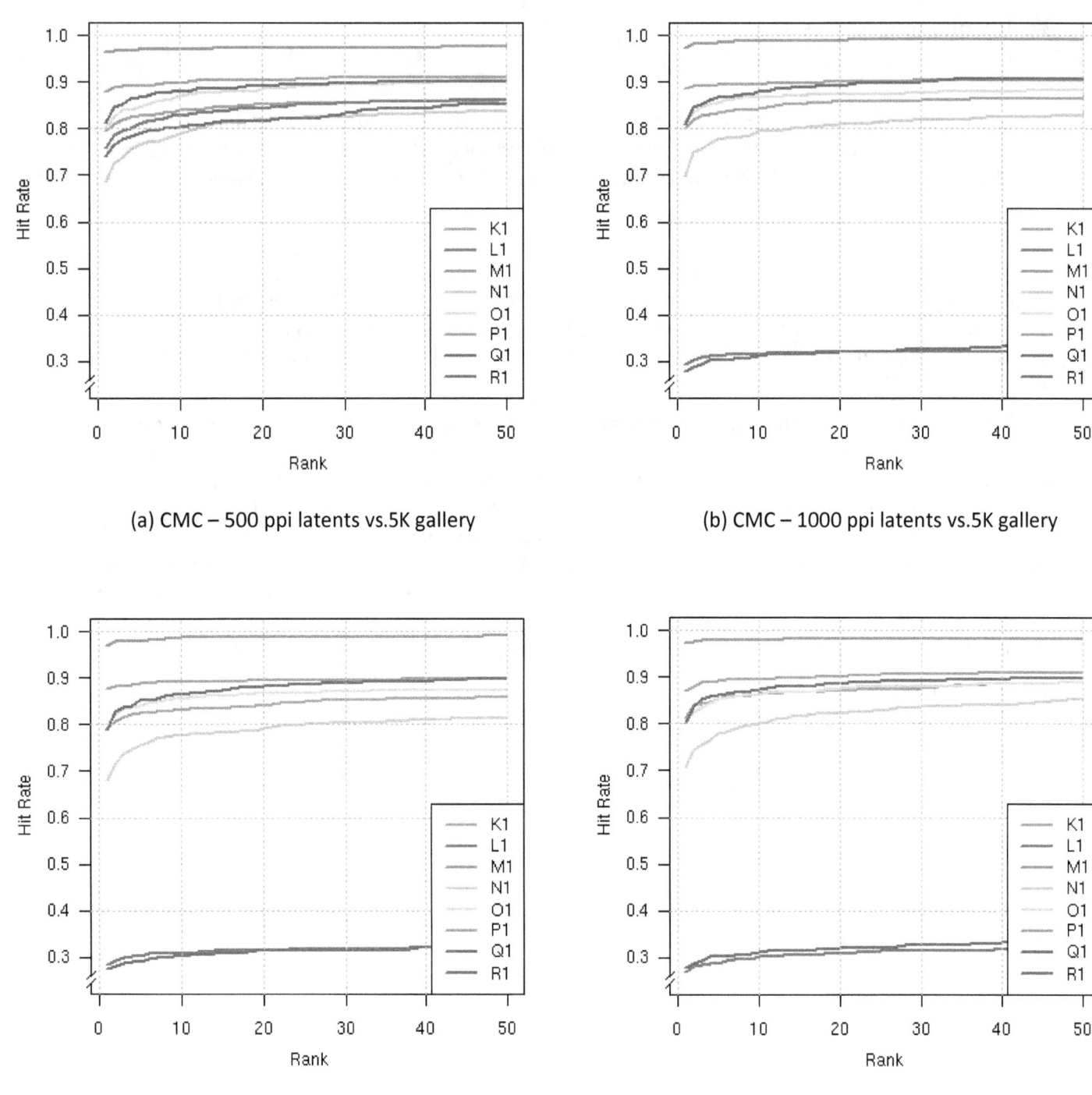

(a) CMC – 500 ppi latents vs.5K gallery

(b) CMC – 1000 ppi latents vs.5K gallery

(c) CMC – 1000 ppi latents vs. 10K gallery

(d) CMC – 1000 ppi latents + ROI vs.5K gallery

Figure 4: CMC of all SDKs for four test cases.
(L1 & R1 have noticeable problems processing 1000 ppi latent images)

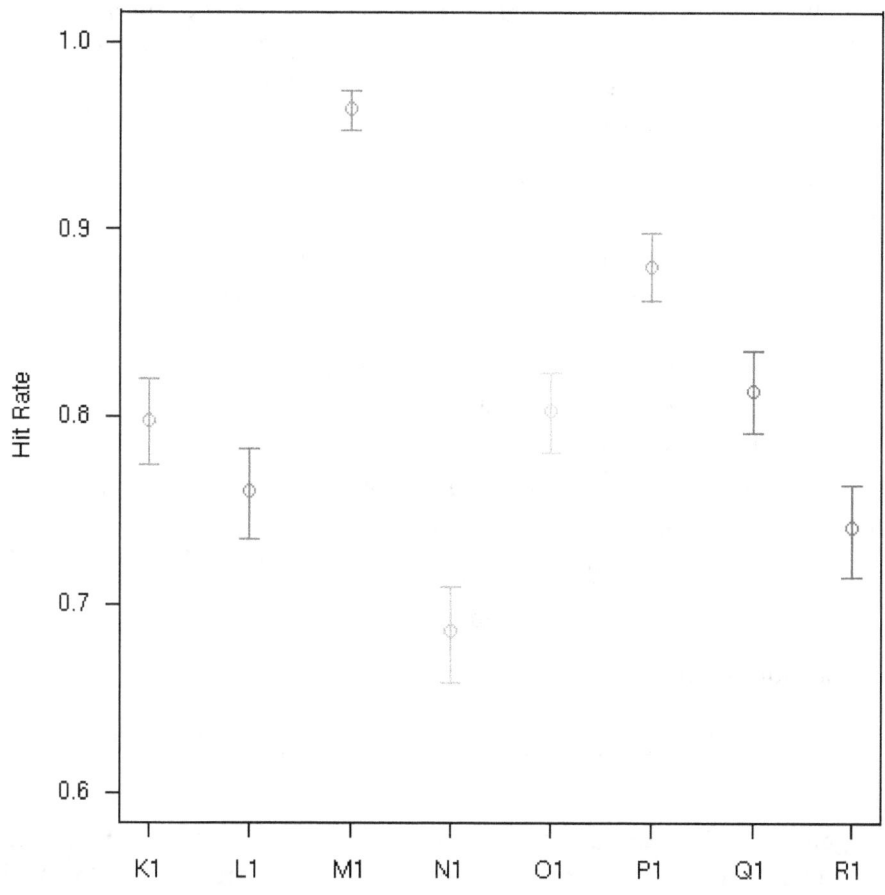

Figure 5: Rank 1 identification rate; 500 ppi latents vs. gallery size 5K with 95% confidence intervals

3.1.2 DET

While CMC is a rank-based identification performance measure, most operational biometric systems report hits and misses relative to a threshold based on matcher score. An analysis of performance relative to threshold is needed here because identification is rarely a closed-universe problem. Instead, in the open-universe case, many latents will not have an enrolled mate. Thus the ideal response to a search is a candidate list of length one for searches that have an enrolled mate and length zero when there is no enrolled mate. In practice, however, there will be more than one entry on a candidate list, because some impostor comparisons will probably exceed threshold, resulting in false matches (Type I error). The opposite case occurs when a genuine comparison score falls below threshold; the result is a false reject, or a miss (Type II error). A Detection Error Trade-off (DET) curve is a plot of Type II error (false reject rate) vs. Type I error (false match rate).

In real-world forensic one-to-many applications, the biometric comparison system returns a candidate list of enrolled identities that may include the correct match. Candidate lists are examined by human experts for the final decision on identity match. There are two types of error that can occur: the correct (genuine) match does not appear on the candidate list, or one or more impostors generate comparison scores higher than the correct (genuine) match and so appear at rank better than the correct match. There are different but closely related terminologies for these two types of errors in the literature, such as False Match and False Non-Match, False Accept and False Reject, False Positive and False Negative, Selectivity and Reliability to name a few.

Absent universally accepted identification error rate metrics, the ISO/IEC 19795 Biometrics Performance Testing and Reporting - Part 1: Framework and Principles suggested new terms. These are False Positive Identification Rate (FPIR) and False Negative Identification Rate (FNIR). The standard, however, leaves the exact mathematical definition to the implementer. So for ELFT we define FPIR as the fraction of candidate lists which contain one or more non-mate entries after the original candidate list has been thresholded at score t and limited to length K. Formally, if

- ID_i denotes subject's i unique ID
- C_i is candidate list for latent print of subject i (ID_i)
- s_{ij} is the comparison score of latent print of subject i (ID_i) and gallery image of subject j (ID_j) (i.e. s_{ii} is the comparison score of latent and gallery images of the same subject, and likewise s_{ij} is the comparison score of latent and gallery images of different subjects.)
- $rank_j$ is the rank of gallery image of subject j (ID_j) on the C_i
- N is the total number of searches
- t is the specified threshold
- K is the candidate list cut-off (limit)

then the false positive identification rate is:

$$FPIR(t, K) = \frac{|\{ C_i : \exists ID_j \in C_i \text{ and } s_{ij} > t, \text{ length}(C_i) = K, ID_i \neq ID_j \}|}{N}$$

Similarly, the false negative identification rate, as a statement of miss rate, is the fraction of candidate lists for which the enrolled mates do not appear in the top K positions with score greater than threshold, t.

$$FNIR(t, K) = \frac{|\{ C_i : \forall ID_j \; s_{ii} < t \text{ or } rank_i \geq K \}| + |\{ C_i : C_i = \emptyset \}|}{N}$$

The ELFT API requires SDKs to produce a candidate list of length 50. In the DETs that follow we restricted rank to values of 1, 20 and 50, and let the threshold run over the entire range of scores reported by the SDK. The dual application of constraints on rank and threshold constitute a candidate list reduction procedure.

To compute FPIR we used impostor scores from searches of galleries containing no mates (see "G1B" and "G2B" above). This was done to reflect operational reality that many searches do not have mates. It is often assumed that the non-mate scores returned in an identification search will be independent of whether or not the enrollment set includes the mate. However, this is not always the case. ELFT included the execution of searches without an enrolled mate, because false matches are a considerable hazard in large gallery one-to-many biometric identification applications, and because the a priori knowledge that a mate exists would allow performance to be artificially improved.

Figure 6 shows DET curves of all eight SDKs for four different test cases when only the rank 1 entry on the candidate list is considered (i.e. reducing candidate list to length 1). DET curves for rank 20 and 50 are shown in Figure 7 and Figure 8, respectively.

The results observed from the DET curves are similar to the CMC results discussed above. M1 outperforms all the other SDKs, followed by P1. L1 and R1 exhibit lower performance on 1000 ppi than on 500 ppi latents. The very small difference in DET curves of ranks 1, 20, and 50 is consistent with our earlier (CMC curve) observation that SDKs most of the time place the correct match at rank 10 or below. Such information is helpful in choosing an appropriate candidate list length in operation. There are small changes observed between the

DET plots when comparing effect of gallery size, latent image resolution (ignoring L1 &R1), and supplementary ROI. These factors will be analyzed more closely in later sections.

Key Observations:

- M1 outperforms all the other SDKs at all thresholds; the second best performer is P1.
- For 1000 ppi latent and gallery 10K, M1 achieves FNIR = 0.149 at FPIR = 0.01.
- The very small difference in DET curves of ranks 1, 20, and 50 is consistent with our CMC-based results in that SDKs most of the time place the correct match at rank 10 or below.

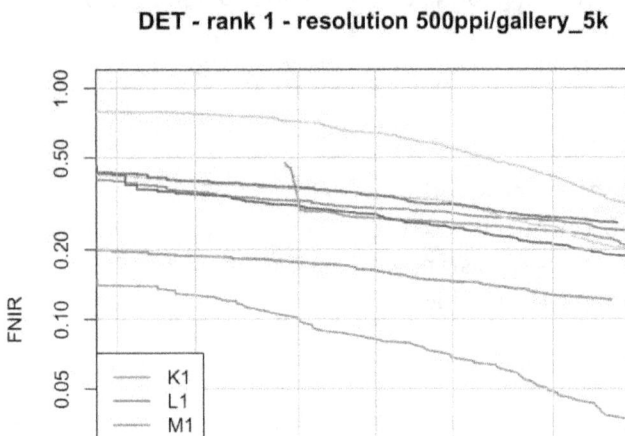
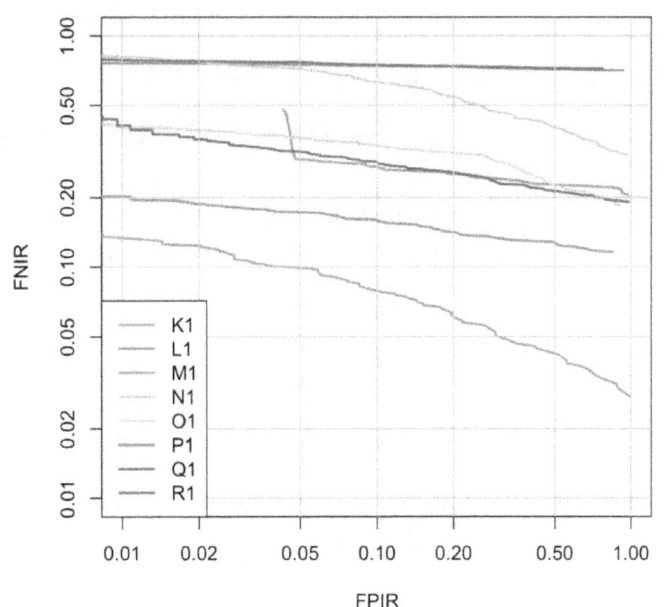
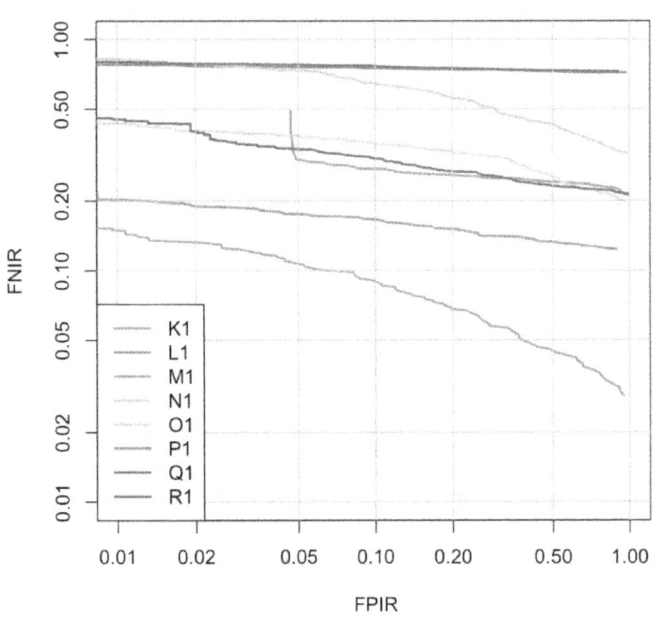
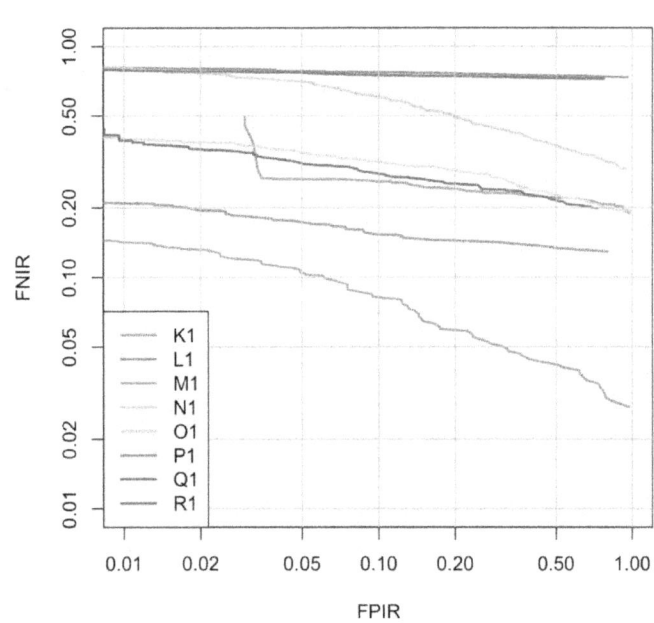

Figure 6: DET **at rank 1** of all SDKs for four test cases

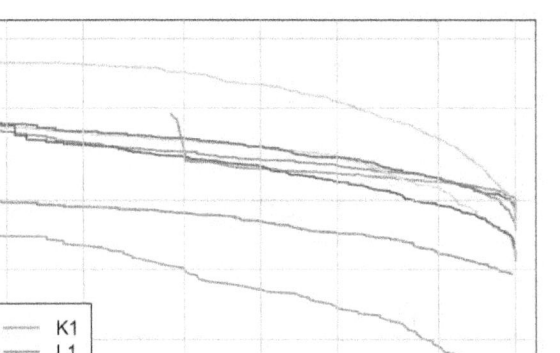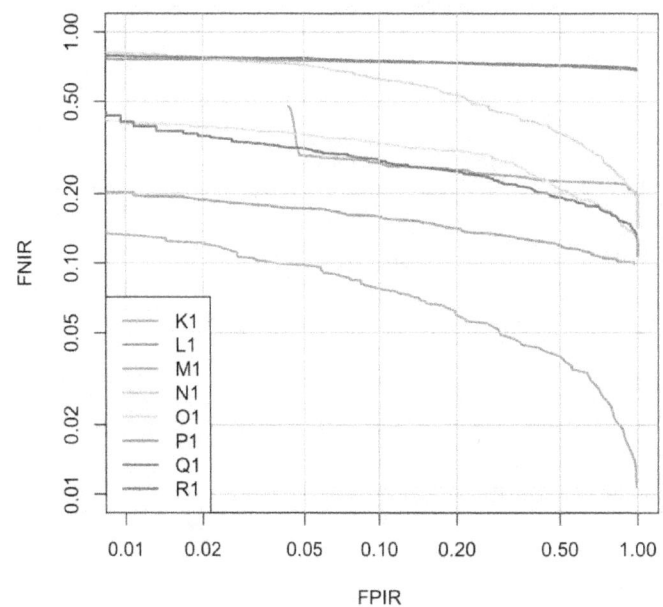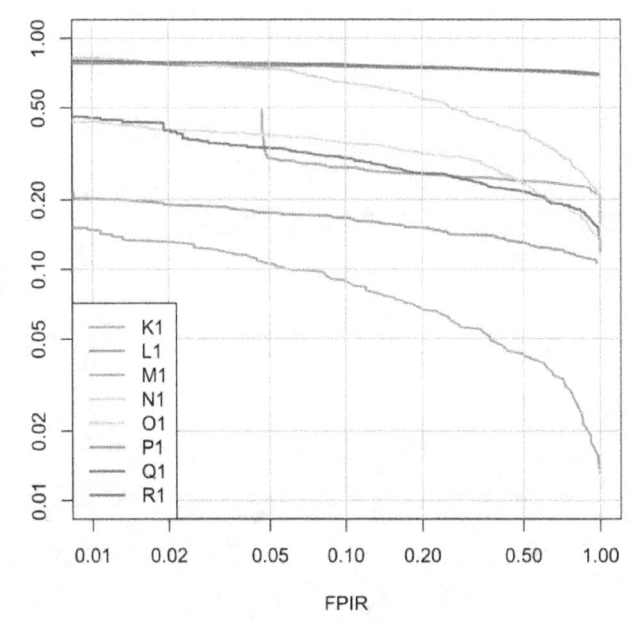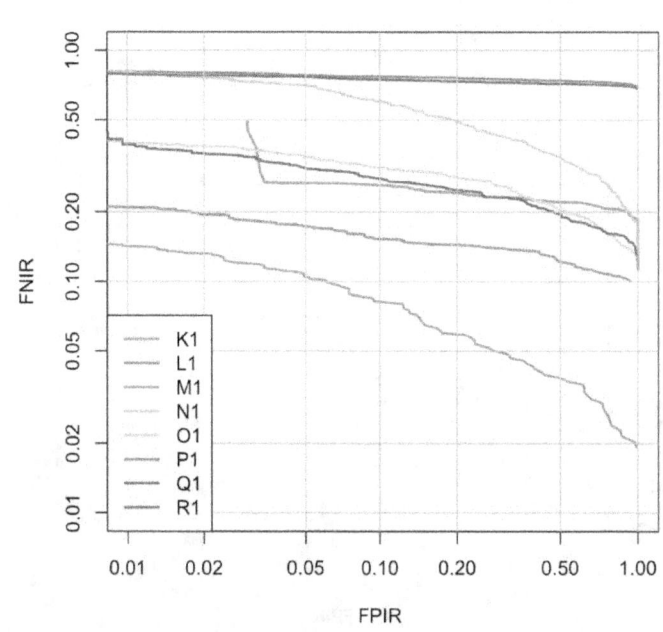

Figure 7: DET **at rank 20** of all SDKs for four test cases

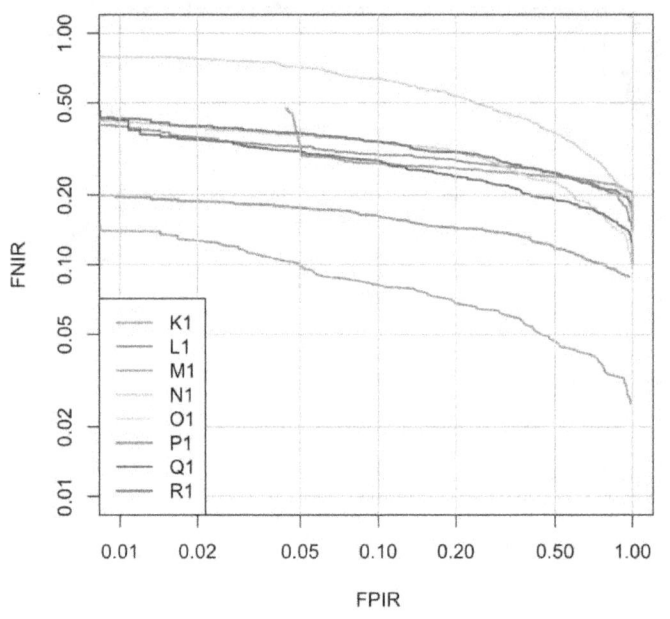
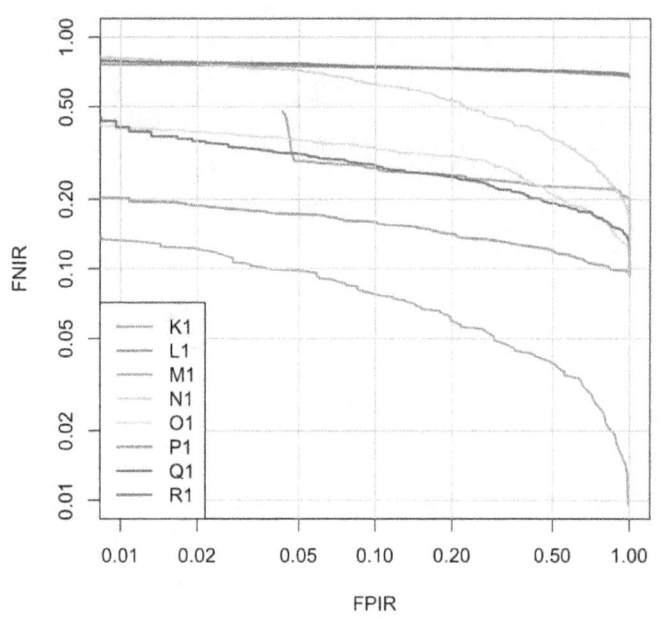
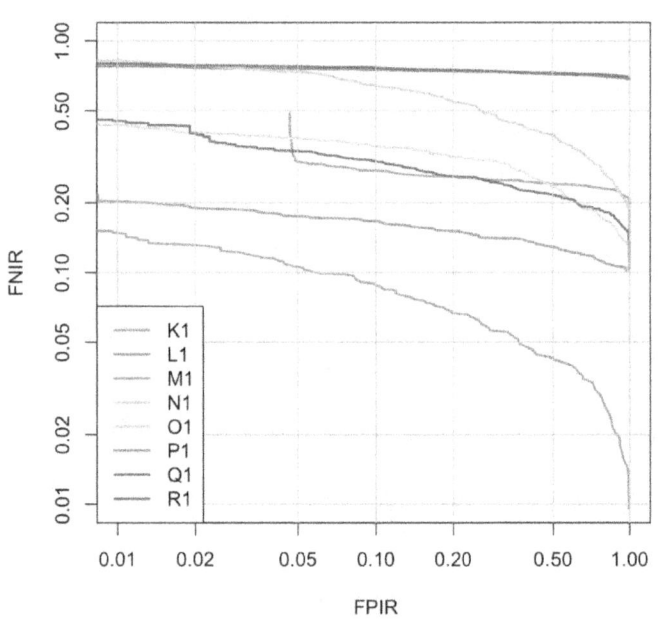
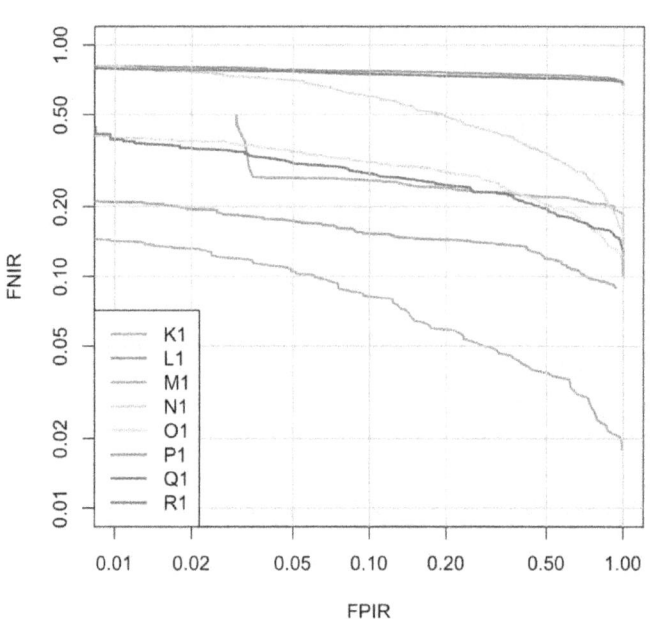

Figure 8: DET **at rank 50** of all SDKs for four test cases

3.2 Comparison with Phase I Results

It is of interest to compare the Phase II performance with the performance using the Phase I dataset. The Phase I dataset was not primarily intended for performance evaluation. Rather, it was intended as a "proof of concept test." As such, it included a large variety of different image types and quality, not necessarily representative of actual case work.

There were 88 latent fingerprint images used to compute performance in the Phase I dataset, and the gallery consisted of 1000 ten-prints. As shown in the table below, when all 15 Phase I SDKs are included in the aggregate, the resulting performance was 59% in first place and 66% on the candidate list. The second row shows the performance for those Phase I SDKs in which the Participant selected to continue to Phase II. The third row shows the performance of revised SDKs submitted for Phase II on the Phase I data. Finally, the fourth row shows the performance of the Phase II SDKs on the Phase II dataset.

Based on Phase I results, NIST expected Phase II performance to be comparable or slightly worse due to the larger gallery. The actual outcome was a surprise, as performance was 12 to 15% higher.

Key Observations:

- On average, performance of Phase II SDKs improved over their corresponding SDKs on the Phase I dataset.
- Phase I dataset is more difficult than the Phase II dataset.

	Performance	
	In first place (top)	On candidate list (1-50)
All Phase I SDKs	59.02%	66.06%
Those going on to Phase II	62.69%	68.37%
Phase II SDKs on Phase I data	68.61%	75.00%
Using Phase II data	80.50%	89.13%

Table 11: Comparison with Phase I results

3.3 Effect of Gallery Size

It is known that as gallery size increases (i.e. the number of individuals in the gallery set increases) the identification rate (as shown on the Y-axis of the CMC curves) decreases [11]. This is known as the scalability problem. We examined the effect of gallery size on performance by comparing the performance of searching the 5K gallery against that of the 10K gallery.

Figure 9 compares identification rate for the two different gallery sizes. Specifically, it shows the rank 1 identification rate of each SDK along with their 95% confidence interval for both the 5K gallery and 10K gallery (an increase in gallery size by a factor of two.) The observed average decrease is 1%.

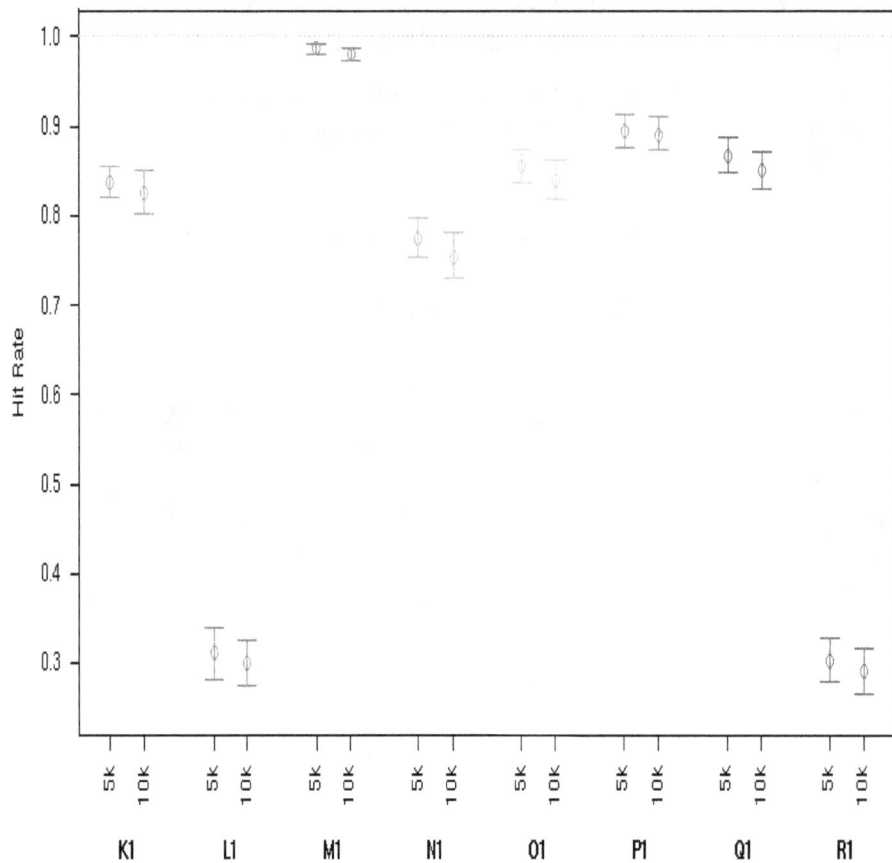

Figure 9: Comparison of Rank 1 identification rate of 1000 ppi latents at two gallery sizes, 5K and 10K; superimposed are 95% confidence intervals

It is interesting to note that the theoretical model described in Appendix A similarly predicts a 0.95% decrease in performance between the 5K gallery and the 10K gallery. The graph below takes this model and predicts the estimated decrease in performance when scaling the gallery size from 5K to the number of subjects plotted along the x-axis. (Note that the x-axis is on the log scale.) The observed average decrease in Phase II testing between the 5K and 10K gallery is plotted as the red point. The model estimates a decrease of performance of just over 10% when scaling from a 5K gallery to a 10M gallery.

The reader is strongly encouraged to not put too much weight in these results. In this case a very simplistic model was used based on the average performance of the eight SDKs in Phase II. The ability of the model to predict the change in performance between the 5K and 10K gallery is promising, but there is no evidence that the model's estimates should be trusted further out. More work in this area is warranted.

Key Observations:

- There is a clear trend that for all SDKs, the Rank 1 identification rate decreases when gallery size is increased from 5k to 10K.
- The amount of change in identification rate is not the same for all SDKs.
- Initial work in modelling the effect of increasing the gallery size has been demonstrated.

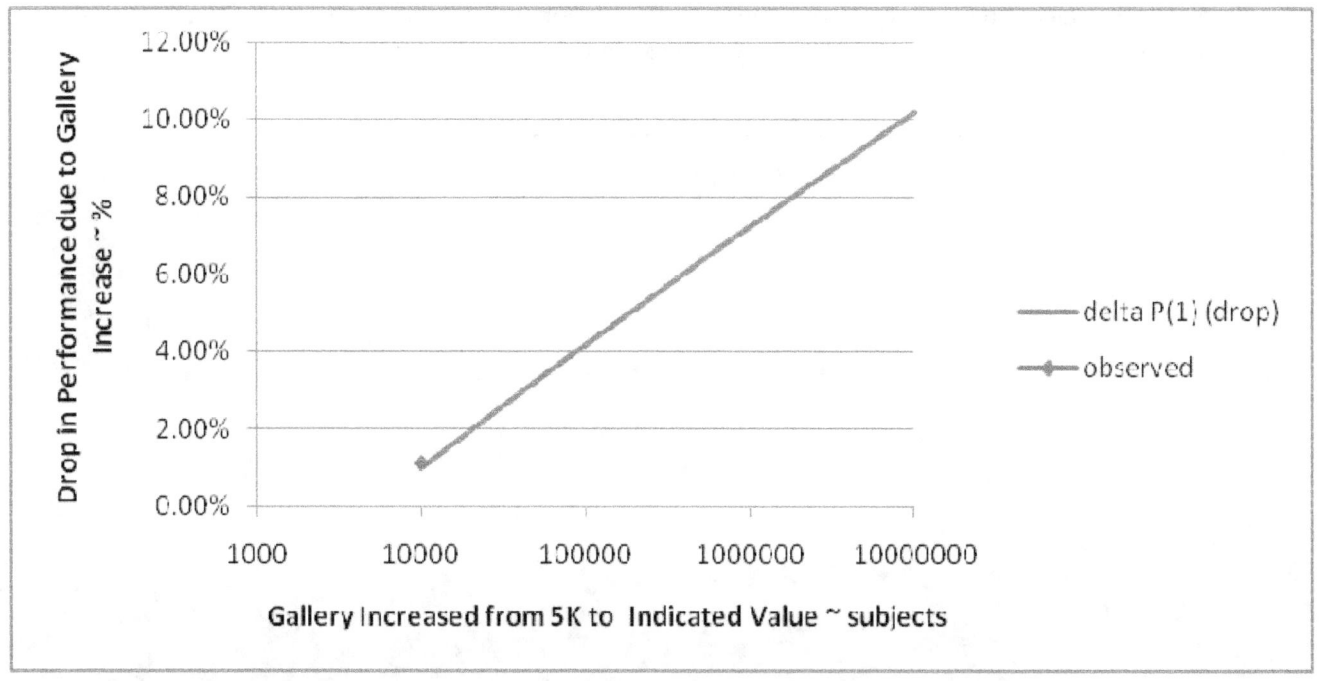

Figure 10: **Estimated average drop in performance due to scaling gallery size.**
Red data point is observed decrease between gallery size of 5K and 10K.

3.4 Effect of Resolution

Increased latent image resolution provides more information, and in principle, should result in better performance. One of the objectives of this study was to analyze the effect of latent image resolution. This is accomplished by comparing searches at 1000 ppi with those at 500 ppi. In both cases, the gallery consists of 5K ten-prints, and note that the ten-print images were scanned at 500 ppi and WSQ compressed. There is a caveat with the latent images: the 500 ppi images were actually scanned at 1000 ppi, and then sub-sampled to 500 ppi. This may produce higher-quality imagery and therefore more favorable results than if the images were originally scanned directly at 500 ppi.

Figure 11 compares identification rate at the two resolutions. Specifically, the figure shows rank 1 identification rate of each SDK along with their 95% confidence interval at 500 ppi and 1000 ppi. SDKs L1 and R1 curiously had much lower identification rates for 1000 ppi latent images than for 500 ppi. Additionally it is noted that Q1's performance slightly drops from 500 ppi to 1000 ppi. The other five SDKs (K1, M1, N1, O1, & P1) all show a small improvement when going from 500 ppi to 1000 ppi, and their average rank 1 identification rate increases by 0.93%. The fact that differences in performance between pairs (same SDK at 500 ppi vs. 1000 ppi) are often smaller than half the confidence interval does not necessarily imply statistically insignificance.

Keeping in mind that searches of a latent at 500 ppi and a 100 ppi are in no way independent observations, the issue of whether higher resolution offers higher accuracy is more completely addressed by counting the occurrences of improved and degraded accuracy. These outcomes are shown in Figure 12.

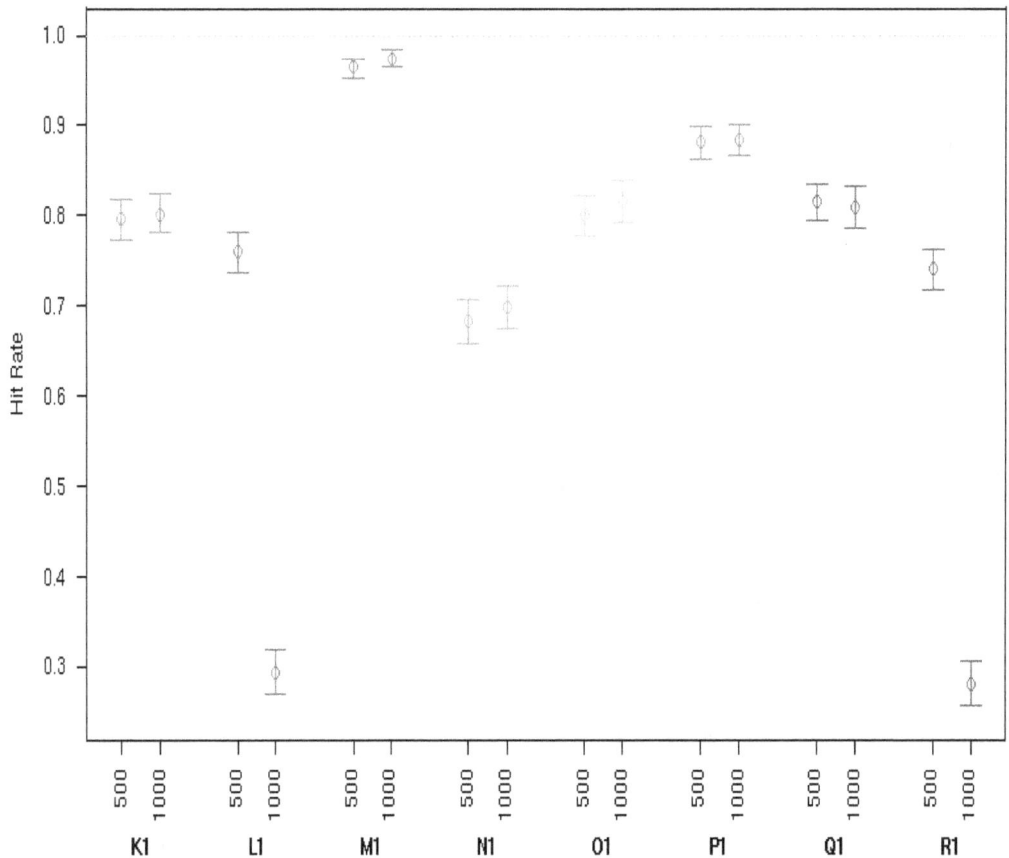

Figure 11: Comparison of Rank 1 identification rate at different resolutions; 500 ppi and 1000 ppi latents vs. gallery size 5K

The following figures provide a detailed comparison between the performance when searching latent images at both 500 and 1000 ppi resolution. (R1 and L1 are excluded.) The general case is that the use of 1000 ppi over 500 ppi causes some hits to be gained, but also some to be lost. M1, Q1 and K1 demonstrate a net benefit, while N1, O1 and P1 show a loss. While M1 realizes the most consistent improvement, all SDKs exhibit degraded rank for some latents. While the changes in the numbers of hits is generally less than 2 percent of the total searches, the net gain for M1 represents a large fraction of the number of latents not hit at rank 50.

Key Observations:

- Excluding L1, R1 & Q1, an average improvement of 0.93% in rank 1 identification rate is observed.
- Increasing resolution from 500 ppi to 1000 ppi causes some hits to be gained but also some to be lost. The net outcome differs between each SDK.

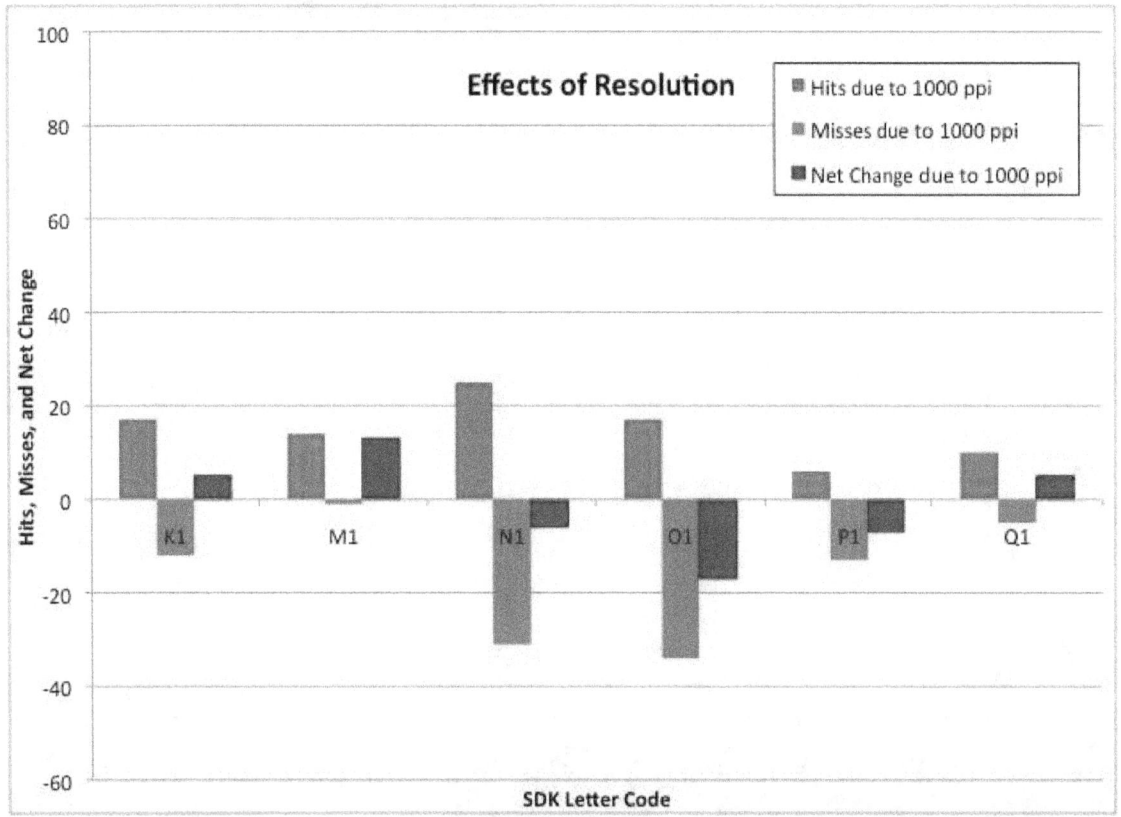

Figure 12: Effect of latent resolution on Rank 50 detection rate; searches of 1000 and 500 ppi latents against gallery of size 5K

3.5 Effect of Region of Interest (ROI) Markup

ROI markup is discussed in section 2.2.6. It should be noted that the instructions to the test participants did not specify that only the ROI image (i.e., after application of the mask) be used. In principle, it was permitted for an SDK to use the original image along with the ROI-extracted image. NIST has no information regarding the actual approach used by any SDK.

Figure 13 shows the rank1 identification rate of the 1000 ppi latents with and without ROI markup, for gallery size of 5K, along with their 95% confidence interval. The results are mixed and show that augmenting the search with ROI markup only slightly improved the performance of matchers K1, N1 & R1, while slightly degraded that of L1, O1, and P1. There is no affect on the performance of M1.

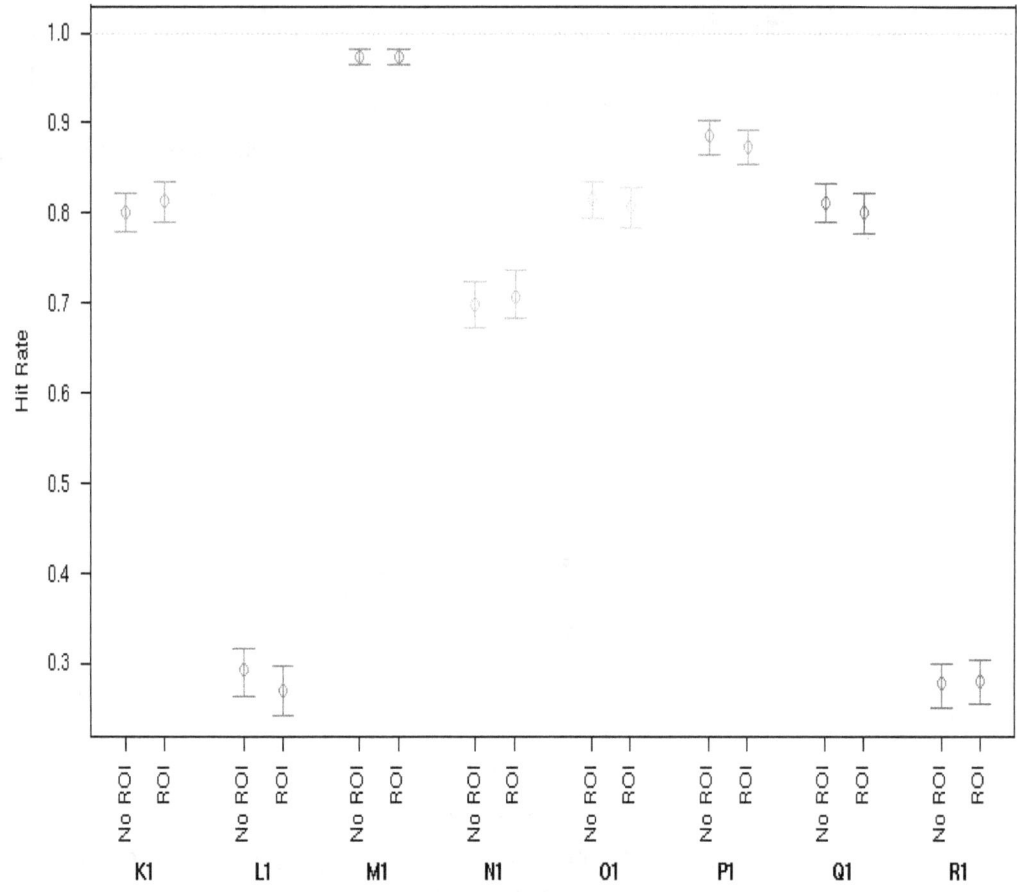

Figure 13: Comparison of Rank 1 identification rate of 1000 ppi latents with and without ROI masking; gallery size 5K; superimposed are the 95% confidence intervals

The following analysis provides additional comparisons between the performance when using the ROIs and not using ROIs. Figure 14 shows the number of cases where the ROI improved the result, the number of cases where it made it worse, and the net difference between these two. Again, the results are mixed. The general case is that the use of ROI causes some hits to be gained but also some to be lost. While K1 and N1 realize the most consistent improvement, all SDKs exhibit degraded rank for some latents. It is interesting to note however, that if ROIs could somehow be selectively applied only to those images where beneficial (i.e., eliminating the cases of degraded performance), it could have resulted in a 3.3% to 4.5% performance increase for three of the SDKs.

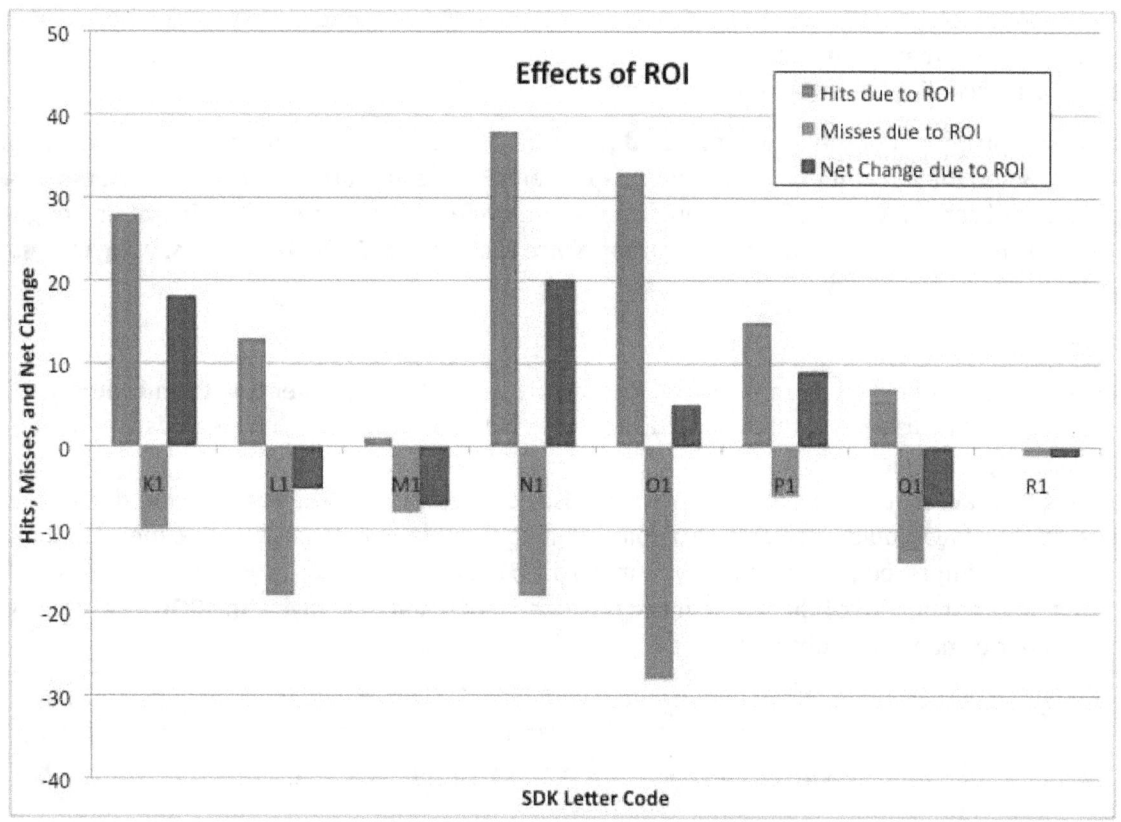

Figure 14: Effect of ROI on Rank 50 detection rate; for searches of 1000 ppi latents with and without ROI against gallery of size 5K

To gain more insight, we looked at how performance was affected by how much of the image was excised. Five "bins" were selected. A latent case was assigned to a bin depending upon the amount of area that was excised from the image when applying the ROI. Bin #1 contains cases where the entire image was used with no area excised; bin #2 contains cases of area excised between 0% and 15%; bin #3 contains cases of area excised between 15% and 35%; bin #4 contains cases of area excised between 35% and 50%; and bin #5 contains cases of area excised greater than 50%. The largest area of excision in the dataset was 80%. The following figure shows the total change in performance for each bin. A "delta-score" was computed based on weighted rankings. Thus a candidate which went from first position (with no ROI) to second position (when using ROI) would receive a delta score of – ½, while one that went from third place to second place would receive + 1/6. The delta performance reported in the figure represents the sum of all delta scores averaged across the SDKs (excluding L1 & R1) for each search.

When only a small amount of the image is excised (0%-15%), the results with ROI are slightly worse than without ROI. When the amount excised is between 15% and 35% results are rather neutral. However, a definite trend is observed when larger amounts of area are excised. For ROIs resulting in excision of greater than 50% of the image, a 3% improvement in performance is observed. So in these cases, using ROI appears to have benefit.

Key Observations:

- The results were mixed for rank1 identification rate of the 1000 ppi latents with and without ROI markup. When using ROI, three SDKs slightly improved, three SDKs slightly degraded, with one SDK remaining the same.
- Using ROI causes some hits to be gained, but also some to be lost. The net outcome differs between each SDK. If ROI could be selectively applied only to those latent images where beneficial, three SDKs could have improved performance by as much as 3% to 4%.
- When more than 50% of the latent image is excised as a result of applying the ROI, an average gain of 3% in performance was observed.

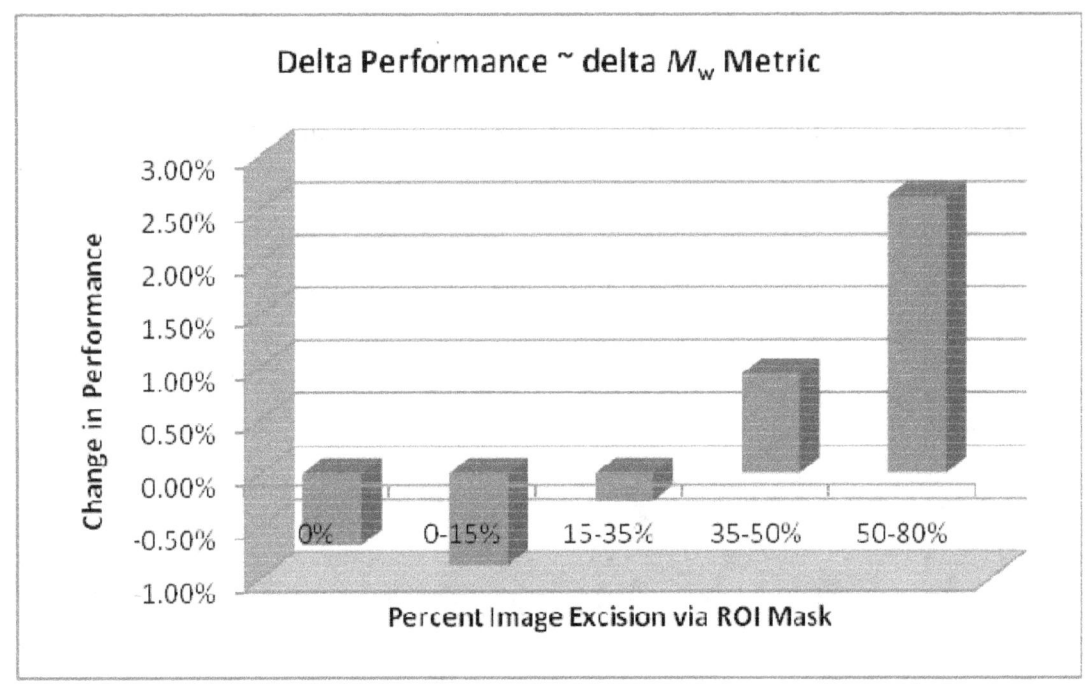

Figure 15: Change in Performance vs. percentage of latent image area marked as ROI.

3.6 Effect of Minutia Count

Each latent has been evaluated by examiners, and information such as the number of minutia points, pattern class, etc., was encoded in the original Type 9 record that was used as part of the transaction (LFFS - latent fingerprint feature search) when submitting to IAFIS. To our knowledge, only one examiner has marked each latent, and given that many aspects of latent markup are subjective, this information might be prone to variation. Nevertheless, such information is useful to study how performance is affected by number of minutiae or pattern class of latent images.

Figure 16 shows how each SDK performed based on the minutia count in each latent searched. The figure reports the distribution of minutia count for latent cases identified at rank 1, rank 2 to 50, and latent cases that were missed. The results were obtained when searching with latents at 1000 ppi on a gallery of 5K. Note that ELFT API required SDKs to produce a candidate list of length 50. Latent cases identified at rank 2-50 are those whose mate appeared on the candidate list, but not as the most probable candidate. Missed latent cases are those that either the mate did not appear on the candidate list or latent cases for which the SDK failed to return a candidate list.

In the Phase II dataset, the latent with the least minutiae has 8, and the latent with most minutiae has 90. The dotted horizontal line in the figure marks the median minutia count at 22 for the latent cases identified at rank one. Furthermore, median minutia count for latent cases identified at rank 2-50 (i.e., appeared in the candidate list, but not at rank 1) is relatively similar to median minutia count for missed latent cases (mate not reported in the candidate list), with their distributions considerably overlapping. This suggests that factors other than minutiae count are needed to distinguish lower-ranking latent cases from missed cases.

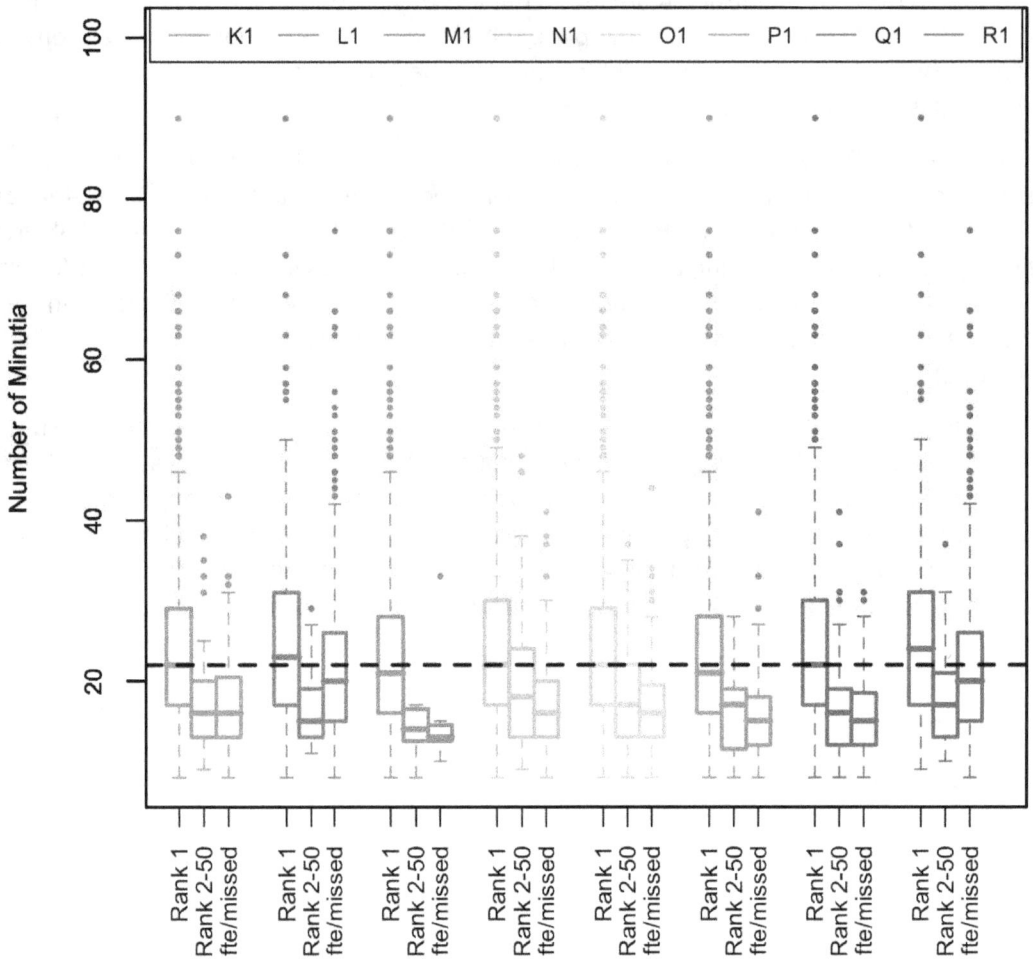

Figure 16: **Box plot of minutia count vs. performance** for latent resolution 1000ppi and gallery size of 5k. (Minutia count was performed manually by fingerprint examiners.)

To determine the extent to which the overall observed detection performance is **affected by minutiae count**, the latent images were divided into three approximately equal bins. **The first bin has the highest minutiae count (from 25 up to 90); the second has medium minutiae count (from 17 to 24); and the third had the lowest minutiae count (8 to 16). The mean minutiae count for the three bins is 35.0, 20.6, and 13.7.** The mean for all cases is 23.

The average performance for the eight SDKs was then evaluated for each bin. **The results are reported in the following figure using latents at 500 ppi on a 5K gallery. A weighted rank-based metric (M_w) is used, which makes allowance for hits in other positions than rank 1. The weight assigned to a hit is the reciprocal of its position on the candidate list.** Thus a hit in first place receives full value (w=1), a hit in second place receives half value (w=1/2), **a hit in third place receives** one third value (w=1/3), and so forth. See **Appendix** A (section A.1) for more discussion on this and other rank-based metrics.

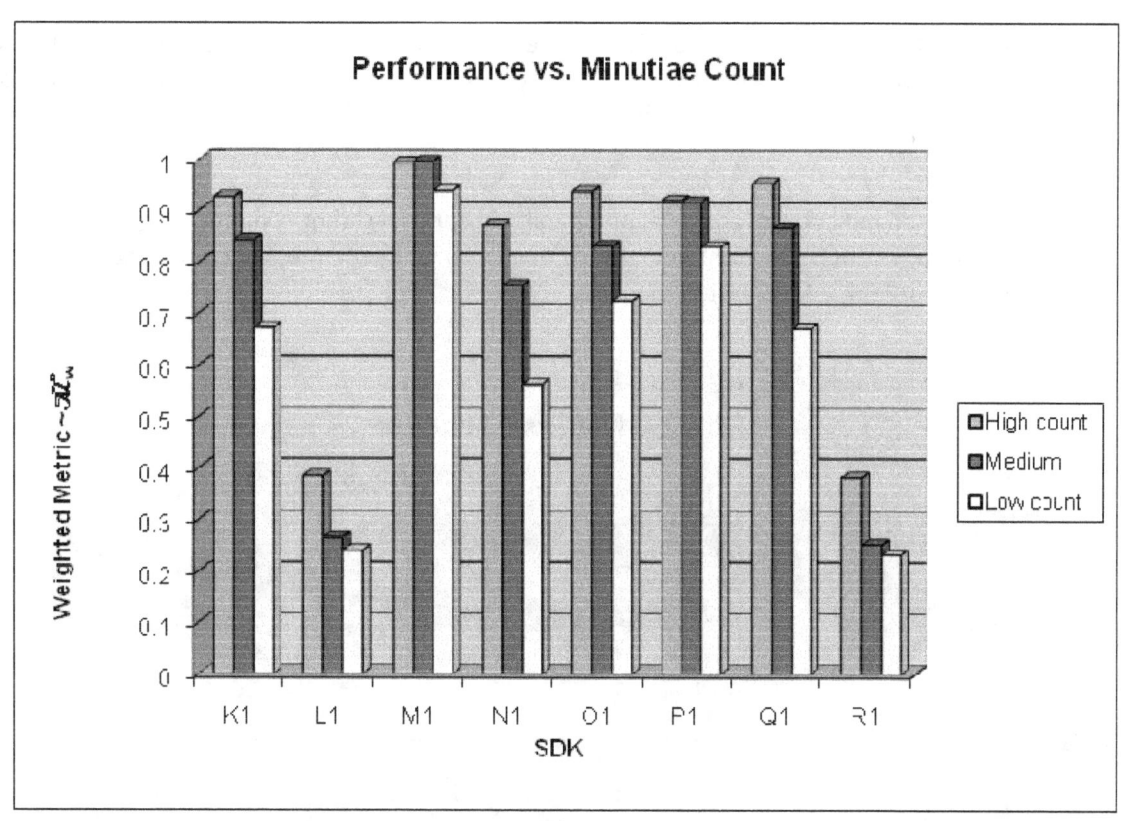

Figure 17: Matcher performance by minutia count category (high, medium and low). Performance computed using average of the M_w metric.

It is clear that performance generally drops off for lower minutia count (successive bins), as expected. In the case of M1 and P1, the performance for the "high" count and "medium" count are almost the same. This suggests that beyond some critical number of minutiae, additional minutiae do not necessarily aid performance for these particular SDKs.

The following graph shows "aggregate" performance plotted against the number of minutiae, where the performance of all eight SDKs was averaged together. There are now four bins. They are similar to the three used above, but now include a "very low minutiae count" bin, so as to better determine the behavior.

Superimposed on the graph is a "trend line." The equation of the trend line (a quadratic) is also exhibited. This trend line may be used to compute a "sensitivity coefficient" (= slope of the trend line). For example, at 20 minutiae, the slope is 1.49% per minutia. This can be interpreted that, if the minutiae count is reduced by one, the M_w statistic can be expected to drop by 1.5%. (The quadratic used here is illustrative of model fitting, but not necessarily the best equation to apply in this case, as it behaves with an inflection point rather than asymptotic as greater numbers of minutiae within latents are encountered.)

The average minutiae count for the Phase II dataset is 23. This is higher than the average minutiae count observed in latent case work, which has been reported to be 17 (this data was provided by an expert latent examiner). The difference in performance between the two minutiae counts can be estimated as 6 x 1.5% = 9%. (E.g., the M_w metric is 71% for 23 minutiae, and 62% for 17 minutiae.) This can be verified directly from the graph.

As expected, rank one identification rate is positively correlated with number of minutiae in a latent print. This implies minutia count could be a proxy for level of difficulty identifying a latent; the more minutia points in a latent print, the "easier" it is to identify the latent.

Key Observations:

- Minutia count is an indicator of the level of difficulty in correctly identifying the latent. The median minutia count for latent cases identified at rank one is 22, while the median minutia count for latent cases identified at rank 2 to 50 is less than 22.

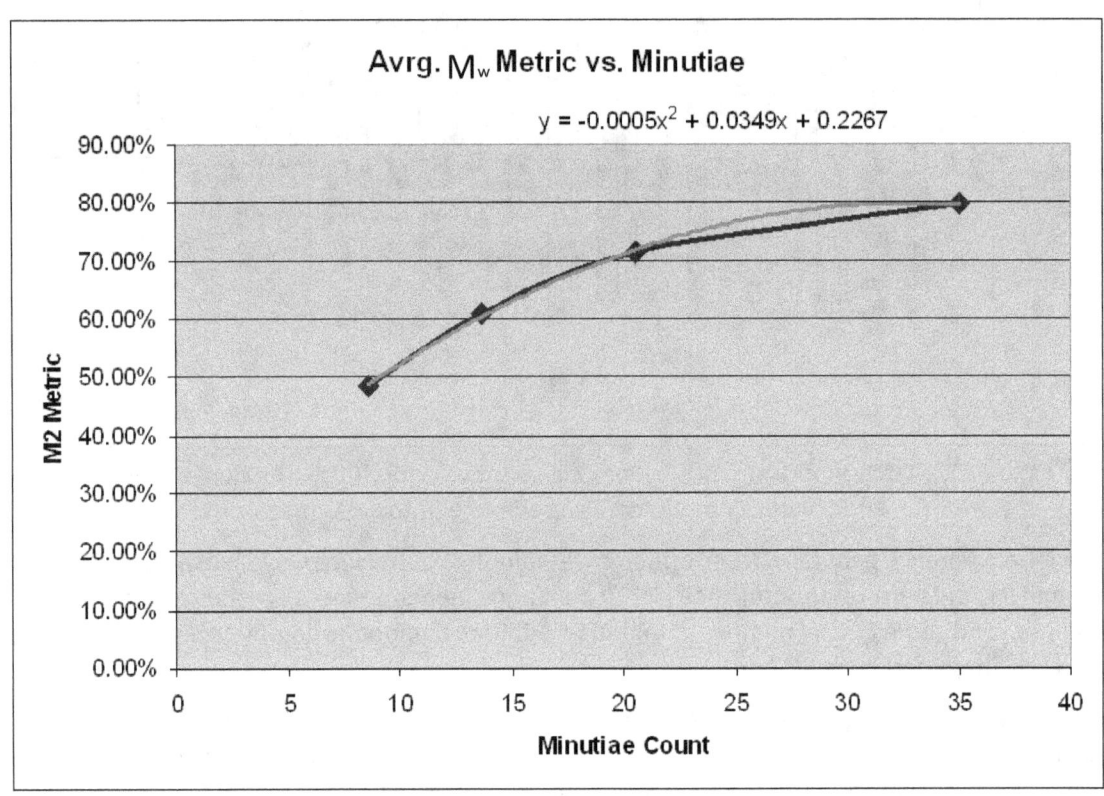

Figure 18: Aggregate matcher performance (M_w) vs. number of minutiae

3.7 Effect of Finger Position

In this section we examine the effect of finger position on performance. Finger position refers to the specific finger of a subject, and is a number from 1 – 10. Finger #1 is the right thumb, and finger #2 is the right index finger, and so forth. Similarly, finger #6 is the left thumb, and finger #7 the left index finger, and so forth. Most latent fingerprints come from finger positions 1, 2, 6, and 7; and not surprising, most of the latent images in ELFT test set are thumb and index fingers. There are very few latents of little fingers (position #5 & #10); therefore results for little fingers are presented but not considered reliable. All fingerprint position information presented here originated from the latent fingerprint examiners who conducted the original case work.

Rank 1 identification rate per finger position for each SDK at resolutions of 1000 ppi and gallery size of 10K is given in Table 12. SDK M1 is least affected by finger position while N1 and K1 performance varies by finger position more than for other SDKs. In order to detect trends, Table 13 was created based on the performance results reported in Table 12. The performance numbers for each SDK were ranked within each hand by finger position. (Note that L1 & R1 are omitted from the ranking due to their unusually low performance on 1000 ppi latents, and little fingers (#5 & #10) have been omitted due to small sample representation.) For example, SDK K1 with the right hand achieved highest performance of 0.85 for the right thumb, and rank 1 is recorded in the corresponding cell in Table 13; likewise, K1 with the right hand achieved lowest performance of 0.55 for the right ring finger, and rank 4 is recorded in the corresponding cell in Table 13. The overall results suggest that multi-algorithm fusion based on finger position could improve performance.

Some trends are seen in Table 13. For right hand, the SDKs rather consistently rank their performance in the order of finger position; #1 ranks 1, #2 ranks 2, etc. The right index finger ranks highest down to the right ring finger. Ranks are a bit more blended for the left hand, but trends are still seen. The left index finger has a higher frequency of rank 2, while left middle and left ring fingers trade off ranks 1 & 3. The left ring finger has the highest frequency of rank 4, and is therefore considered the most difficult to match on the left hand.

Figure 19 plots the results of Table 12 in the bottom graph (latents at 1000 ppi and gallery size of 5K). The top graph is for latents at 500 ppi and gallery size of 5K, and is provided for the benefit of SDKs L1 & R1. These graphs support the trends observed above.

Key Observations:

- SDK M1 is least affected by finger position, while N1 and K1 performance varies by finger position more than for other SDKs.
- Looking at just the thumb, index, middle, and ring fingers on each hand, there is some evidence that latent search performance is highest with thumbs, next with index fingers, and lowest with ring fingers. Results on little fingers were not analyzed due to very small sample size representation in the Phase II dataset.

Finger position	1	2	3	4	5	6	7	8	9	10
Number Searches	247	127	64	20	6	147	101	64	48	11
K1	0.850	0.772	0.672	0.55	0.833	0.816	0.782	0.844	0.688	0.818
L1	0.356	0.236	0.219	0.35	0.000	0.306	0.277	0.234	0.208	0.091
M1	0.996	0.945	0.938	1.00	1.000	0.959	0.960	1.000	0.979	1.000
N1	0.789	0.622	0.531	0.50	1.000	0.673	0.733	0.609	0.521	0.545
O1	0.854	0.795	0.766	0.75	1.000	0.796	0.743	0.828	0.688	0.727
P1	0.891	0.890	0.875	0.70	0.833	0.876	0.891	0.860	0.833	1.000
Q1	0.866	0.732	0.734	0.75	0.833	0.755	0.802	0.828	0.625	0.818
R1	0.332	0.244	0.250	0.30	0.000	0.327	0.248	0.219	0.146	0.091

Table 12: Rank 1 identification rate per finger position for latent searches at 1000 ppi and gallery size of 10K. Finger position for each latent in the probe set was determined manually by fingerprint examiners. Number of latent images with each finger position is shown in the second row.

Finger position	1	2	3	4	5	6	7	8	9	10
K1	1	2	3	4		2	3	1	4	
M1	2	3	4	1		4	3	1	2	
N1	1	2	3	4		2	1	3	4	
O1	1	2	3	4		2	3	1	4	
P1	1	2	3	4		2	1	3	4	
Q1	1	4	3	2		3	2	1	4	

Table 13: Ranking within each SDK, with each hand, by finger position based on performance reported in Table 12. (SDK L1 & R1 along with little fingers (#5 & #10) are omitted.)

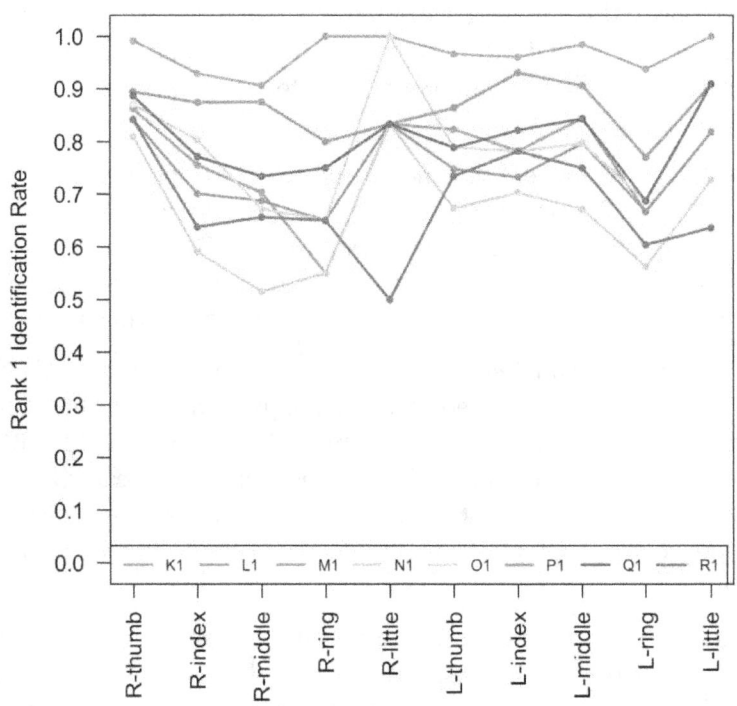

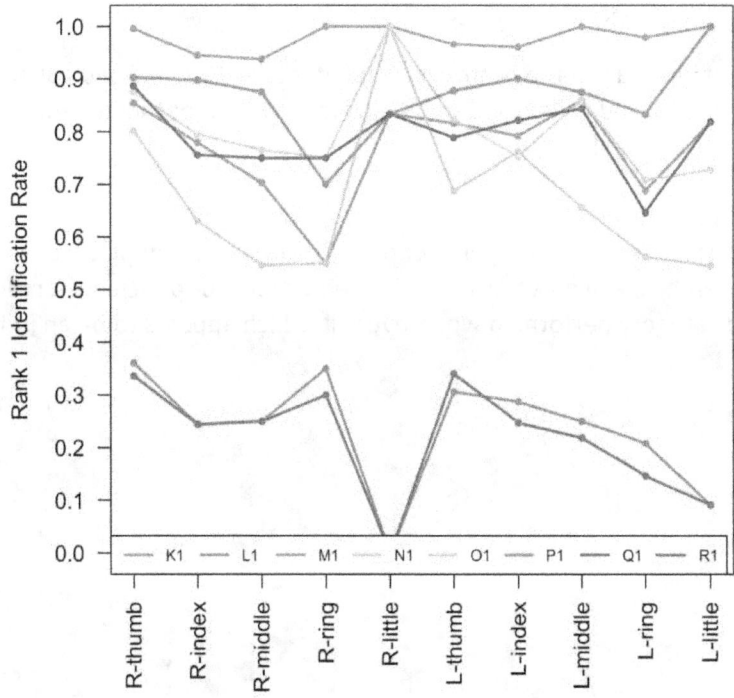

Figure 19: **Rank 1 identification rate** for all SDKs per latent finger position for two test cases. (Finger position for each latent was determined manually by fingerprint examiners.)

3.8 Effect of Pattern Class

NIST investigated how identification rate varies for different latent pattern classifications. Each latent in the probe set has been assigned one of the four basic pattern classes: arch, left slant loop, right slant loop, and whorl; if none of these could be established with certainty, a category of "undetermined" was assigned. All pattern class information presented here originated from the latent fingerprint examiners who conducted the original case work.

About 42% (348 of 835) of ELFT Phase II latent searches are whorls, 23% right loops, 23% left loops, 4% arches, and 8% were undetermined. The FBI's Criminal Master File contains a distribution of 65% loops, 30% whorls, and 5% arches. If this is representative of the US population, then it appears that the Phase II dataset set is somewhat overrepresented in whorls, and underrepresented in loops. This is most likely due to the dataset containing a large percentage of thumbs, as thumbs are more likely to have whorls.

Rank 1 identification rate for each SDK for latent images of 1000 ppi and gallery size of 10K is given in the Table 14. In order to detect trends, Table 15 was created based on the performance results reported in Table 14. The performance numbers for each SDK were ranked by pattern class. (Note that L1 & R1 are omitted from the ranking due to their unusually low performance on 1000 ppi latents.) For example, SDK K1 achieved highest performance of 0.813 on whorls, and rank 1 is recorded in the corresponding cell in Table 15; likewise, K1 achieved lowest performance of 0.742 on the undetermined category, and rank 5 is recorded in the corresponding cell in Table 15.

Some trends are seen in Table 15. In terms of SDK performance, whorls are either ranked first or second, indicating higher matchability. The results for arches are quite bipolar. Four SDKs achieved highest performance on arches, while two SDKs achieved lowest performance on arches. The rankings for loops are mixed, with medium ranks (mostly ranks 3&4.) The undetermined category has the highest frequency of ranks 4 & 5, indicating that they are the most difficult to match. The observation that whorls have higher matchability may be indicative of AFIS-bias in the selection of latent cases, which would explain why whorls are somewhat overrepresented in the Phase II dataset.

Figure 20 plots the results of Table 14 in the bottom graph (latents at 1000 ppi and gallery size of 10K.) The top graph is for latents at 500 ppi and gallery size of 5K, and is provided for the benefit of SDKs L1 & R1. These graphs support the trends observed above.

Key Observations:

- Latent search performance was higher with whorls. The results for arches were bipolar; four SDKs performed best on arches; while two SDKs performed worst. Loops achieved medium performance. The undetermined category performed worst over all, which appears to be an indication of low latent image quality.

	Whorl 348	Arch 30	Right Loop 195	Left Loop 196	Undetermined 66
K1	0.813	0.933	0.754	0.791	0.742
L1	0.322	0.233	0.246	0.270	0.273
M1	0.991	1.000	0.954	0.959	0.955
N1	0.701	0.600	0.672	0.679	0.621
O1	0.807	0.867	0.795	0.791	0.773
P1	0.882	0.933	0.877	0.872	0.848
Q1	0.807	0.700	0.805	0.765	0.742
R1	0.328	0.167	0.221	0.255	0.273

Table 14: Rank 1 identification rate per pattern class for latent searches at 1000 ppi and gallery size of 10K. Pattern class for each latent in the probe set was determined manually by fingerprint examiners. Number of latents with each pattern class is shown in the first row.

	Whorl 348	Arch 30	Right Loop 195	Left Loop 196	Undetermined 66
K1	2	1	4	3	5
M1	2	1	5	4	3
N1	1	5	3	2	4
O1	2	1	3	4	5
P1	2	1	3	4	5
Q1	1	5	2	3	4

Table 15: Ranking within each SDK by latent pattern class.
(SDK L1 & R1 are omitted.)

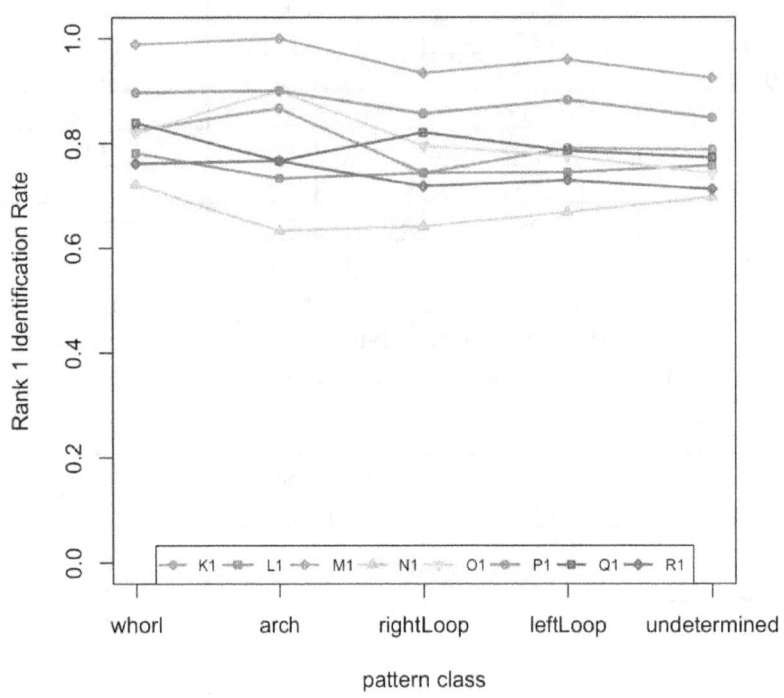

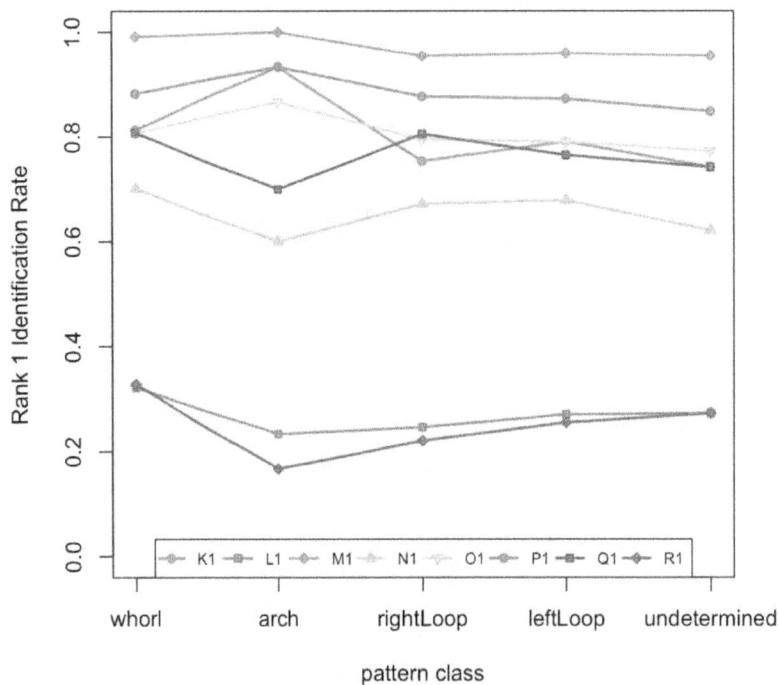

Figure 20: **Rank 1 identification rate** of all SDKs per latent pattern class for two test cases. (Pattern class of each latent was determined manually by fingerprint examiners.)

3.9 Timing Results

The observed execution times for all SDKs tested are presented here as elapsed time (i.e. "wall clock time") measurements. These timings were generated using the system clock on systems executing only a single SDK in addition to the typical system software load. Note that some SDKs implemented individual execution phases as multi-threaded processes, which were able to more fully utilize the four processing "cores" available on each system.

Figure 21 presents the average execution time for enrolling a gallery subject. There appears to be a variation of about a factor of 6 between the different SDKs. The mean enrollment time is approximately 50 seconds, so that to enroll the 10K gallery would require about 5.8 days. Two of the best performing matchers in the test (M1 and Q1) in terms of hit-rate were the slowest in terms of enrollment time. (The rankings shown in Figure 5 are used in this section for the purpose of SDK comparisons. These rankings are based on searching 500 ppi latents against the gallery of 5K.) The lowest performing matcher (N1) was the fastest. Some matchers' enrollment times were affected by gallery size: L1 & Q1 become faster per ten-print for the larger gallery size; P1 becomes slower per ten-print for larger gallery size.

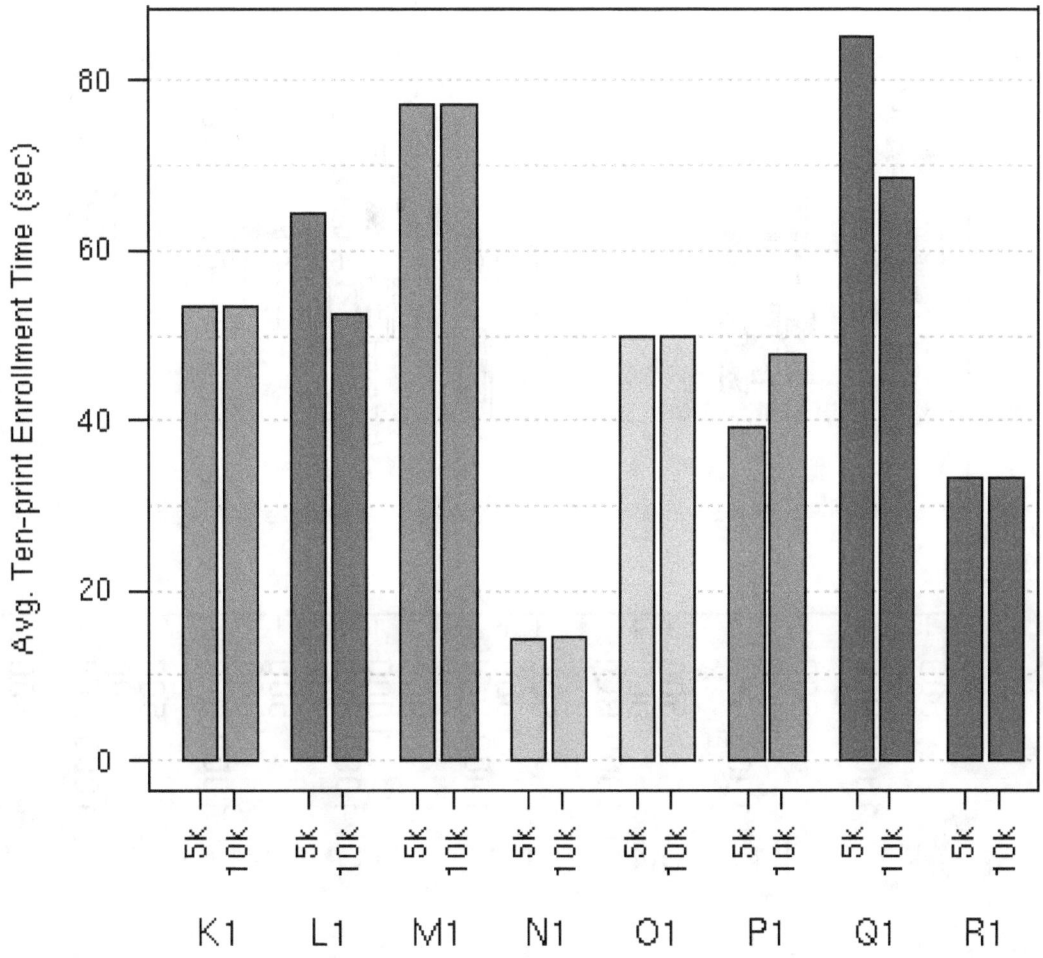

Figure 21: Ten-print enrollment times

Figure 22 shows the enrollment time per latent for three cases: 500 ppi, 1000 ppi, and 1000 ppi with ROI. The median latent enrollment times vary more by SDK than do the ten-print enrollment times; there appears to be at least a factor of ten variation. Three of the highest performing matchers (M1, P1, Q1) in the test are also the slowest; and the best performing matcher (M1) is more than an order of magnitude slower than the next slowest matcher. The lowest performing matcher (N1) was the 2nd fastest. Median times are about equal for both image resolutions (500 &1000 ppi) for all matchers except N1. The addition of ROI has no affect on median times in 2 cases (R1,Q1) ; in four cases (K1,L1,O1,P1) the median time is decreased slightly, and in 2 cases (M1, N1) it increases slightly.

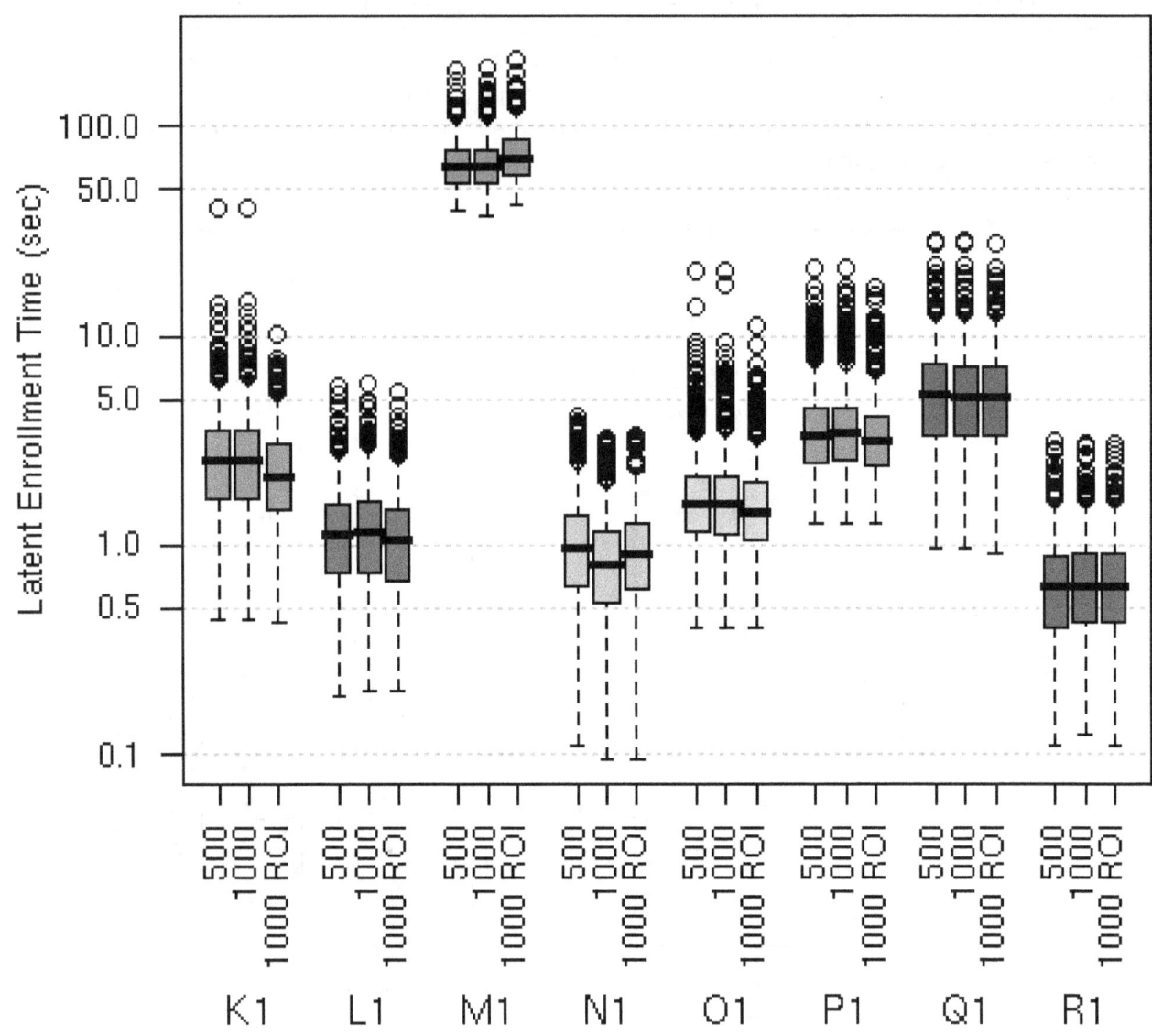

Figure 22: Latent enrollment time

Figure 23 compares total search times. Two cases are shown for each SDK, 5K and 10K gallery. Not surprisingly, the search times for the 10K gallery are very nearly twice those for the 5K gallery. The variation by SDK of individual search times is again large, and appears to be at least a factor of ten. The slowest matcher, K1, was the 5th most accurate performing matcher overall. The second and third slowest matchers, P1 & M1, were the first and second most accurate performing matchers overall. The fastest matcher, N1, was the least accurate matcher overall. The increase in median search time between 5K and 10K gallery was usually linear (i.e. a factor of 2). However, in three cases it increased less than a factor of two (Q1,P1,K1).

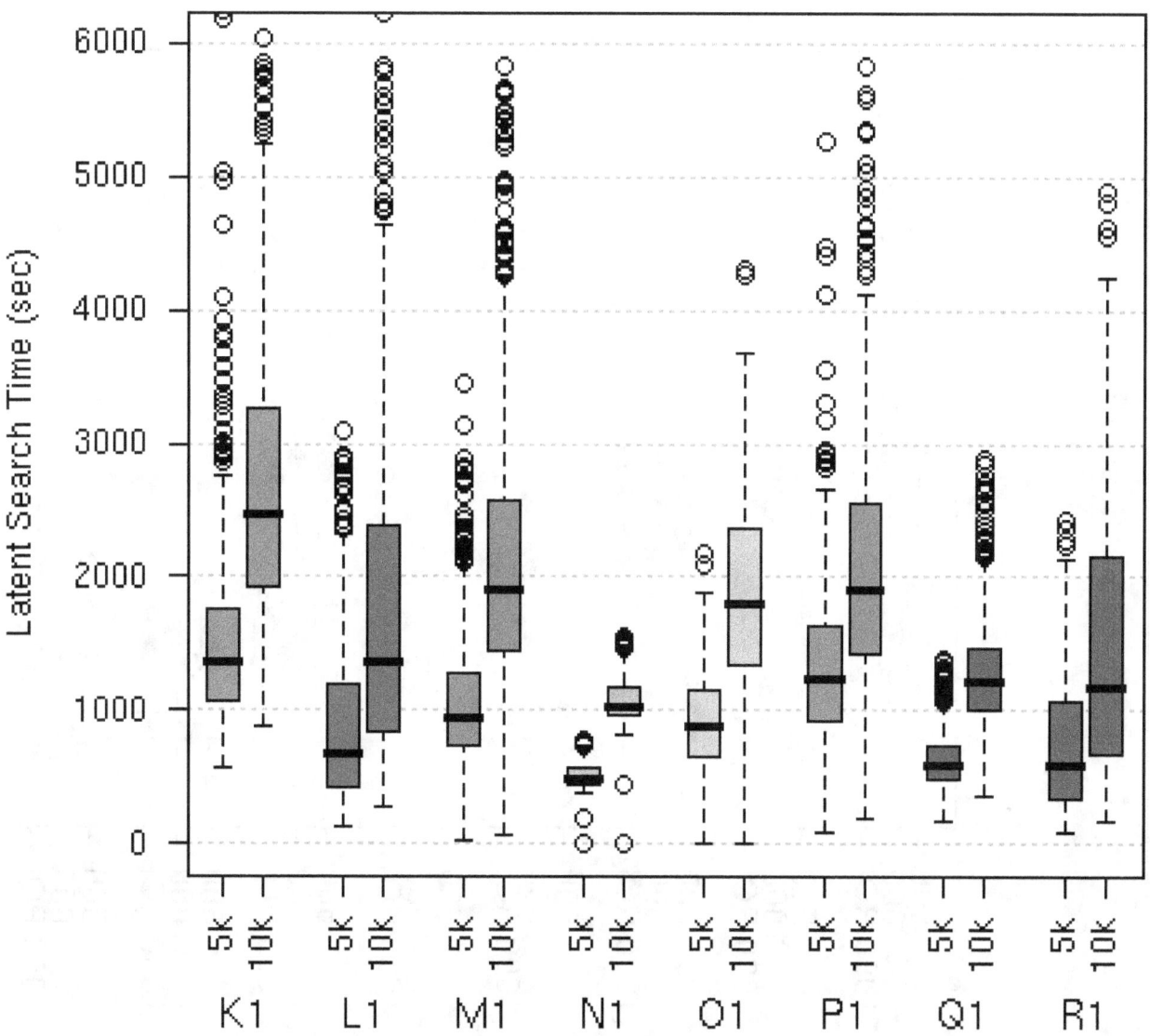

Figure 23: Search times for Gallery sizes 5K and 10K

Figure 24 is similar to the previous one, but breaks the data down by the three cases: 500 ppi, 1000 ppi, and 1000 ppi with ROI. Search times for the three cases appear very similar, though the 2nd case is often the slowest. The slowest matcher was the 5th most accurate performing matcher overall. The second and third slowest matchers were the first and second most accurate performing matchers overall. The fastest matcher was the least accurate overall (ignoring the 1000 ppi cases for L1 & R1). There was a slight increase in median search time between 500 & 1000 ppi across all matchers except O1 (excluding L1 & R1). When going from 1000 ppi to 1000 ppi with ROI, there was a decrease in observed median search time across all matchers except R1 (which had a slight increase). The greatest decrease occurred when ROI was added for L1.

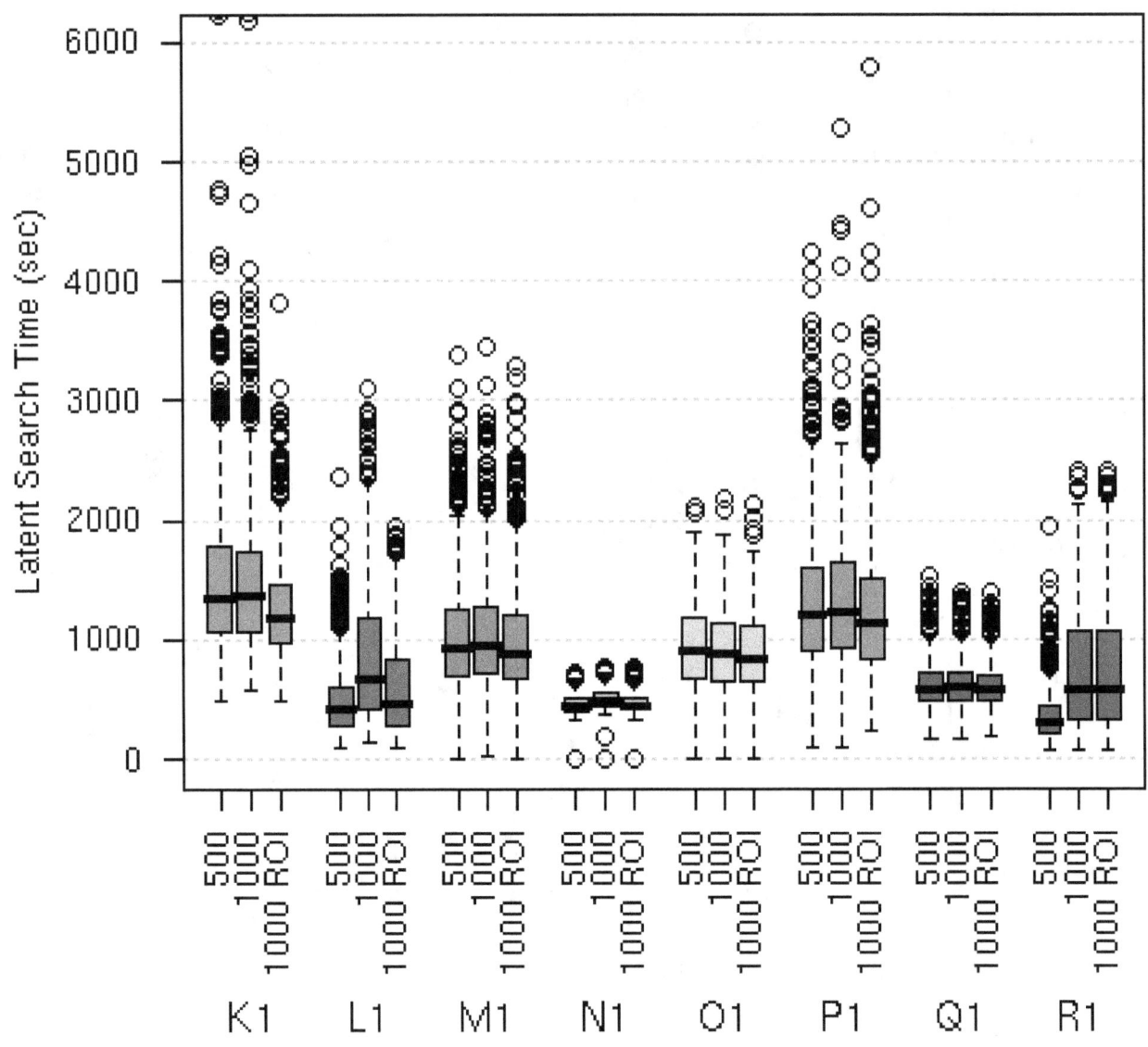

Figure 24: Search times for 500 ppi, 1000 ppi, and 1000 ppi + ROI

Key Observations:

- The most accurate SDKs tend to execute for longer periods of time than less accurate ones during ten-print enrollment, latent feature extraction, and latent search operations. The least accurate SDK nearly always executed for the least amount of time.
- Latent image resolution and use of ROI masks have small, mixed effects on latent feature extraction and search times.
- Increases in gallery size in general (but not always) have proportional increases in latent search times.
- Note that the hardware used in this study may not be representative of operational systems. Therefore, all timing results reported may not be directly comparable to operational scenarios, which may involve different hardware and software.

4 Probability Scores vs. Raw Scores

The ELFT Phase II API specifies that every candidate reported by an SDK should be accompanied by an estimate of the likelihood that the candidate is a true mate of the search latent (called a probability score). The exact manner of calculation was not specified, but some guidelines were provided in the CONOPS. The estimate is to be assigned a value between 0 and 100. A value near 100 is to be interpreted as "this candidate is almost certainly a true mate"; while a value near zero is to be interpreted as "this candidate is very unlikely a true mate."

The motivation behind this was three-fold:

1. The magnitude and range of raw match scores vary greatly between different algorithms from different providers. Some matchers produce scores that are small fractions (say 0.1), while others produce scores in the hundreds of thousands. A latent expert needs to be thoroughly familiar with a particular matcher to assess whether a given raw score is significant or not. Probability scores were introduced into this study in an attempt to simplify this by standardizing the range and meaning of the scores making them more intuitive for human judgment.

2. It is also desirable to have scores that can be used for candidate list reduction. Candidate lists produced by current-generation latent matchers are "cluttered" with many candidates that are patently improbable – for example, when the search latent and the candidate differ in pattern class. A goal of ELFT is to study how to eliminate as many "nuisance" candidates as possible. Using probability scores provides a mechanism whereby candidates having very low likelihood of being a mate (for example 2% or less) could be pruned from the list.

3. Probability scores have the potential to factor in information in addition to just the matcher's raw score including: quality of the latent, number of minutiae found, number of minutiae matched, pattern class, finger position, and the size and difficulty of the gallery. Factoring in this information has the potential to generate more robust scores that place true mates closer to the top of candidate lists, thus increasing performance.

4.1 Reported Probability Scores

All the SDKs reported probability scores in their generated output; however, analysis of the eight SDKs shows a variety of approaches and results. The bar graph in Figure 25 compares the mean value of probability scores for true mates to that of true non-mates (i.e., impostors). L1 and R1 computed probability scores for true mates too low to be useful. The other SDKs computed reasonably good probability estimates for true mates, but their probability estimates for impostors tend to be too high.

Figure 26 shows the ratio of mean probability scores for the two distributions (true mates and imposters) by SDK. The largest ratio (i.e., highest separation) was obtained by M1 and Q1. We may therefore consider these two as having come closest to implementing a workable candidate list reduction measure. However, the mean value of 5 for impostors is considered too high for effective candidate list reduction – a candidate which has a 1/20 chance of being a mate is worth looking at.

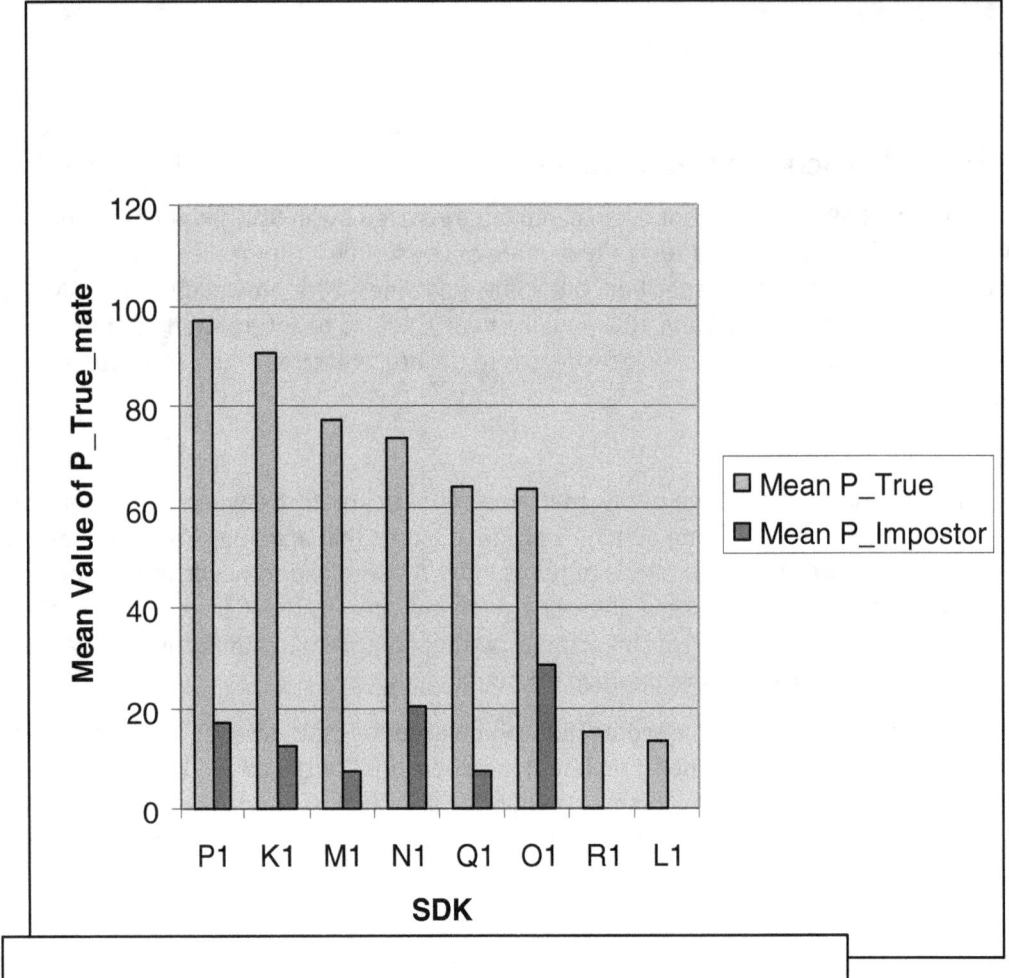

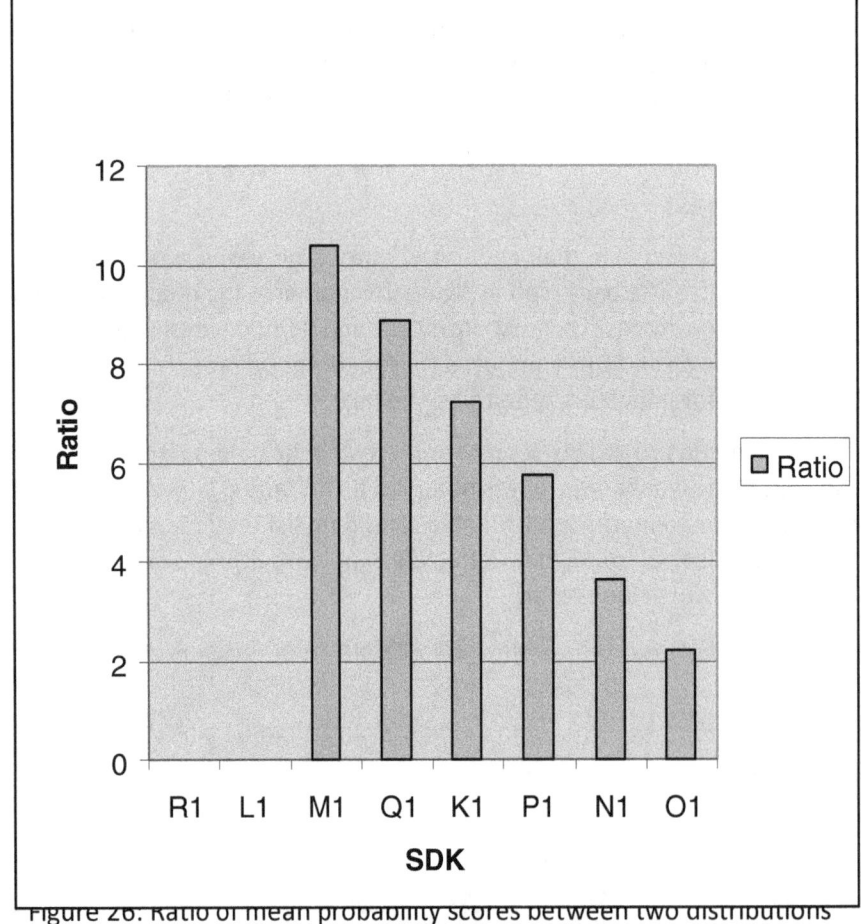

Figure 26: Ratio of mean probability scores between two distributions (true mates and imposters) by SDK

4.2 Re-computing DET Using Probability Scores

The reported probability scores can be substituted in place of raw matcher score when computing DET curves. This may or may not produce the same result, depending upon whether additional information enters into the calculation of the probability scores. Figure 27 compares DET results between using raw matcher scores and using the provided probability scores. The performance of two SDKs (M1 & N1) improves when probability scores are used, O1 shows improvement down to an FPIR around 5%, while no change is noticed from P1, K1, and Q1.

Comparing the difference in performance for SDK M1 between the bottom two graphs (c & d) in the figure (latents @ 1000 ppi on 10K gallery), we see that at an FPIR of 5%, the FNIR with raw scores is 11%, while the corresponding FNIR with probability scores is 6%. An even larger change in performance is seen with SDK N1. These results are promising, but it appears more work is needed in this area.

Key Observations:

- Results were mixed, but promising, with two SDKs clearly demonstrating enhanced capability to reduce false matches.
- More work in the area of probability scores is warranted for the purposes of increasing accuracy, but also in generating probability scores that are more intuitive to human examiners, more effective in candidate list reduction, and more interoperable across SDKs.

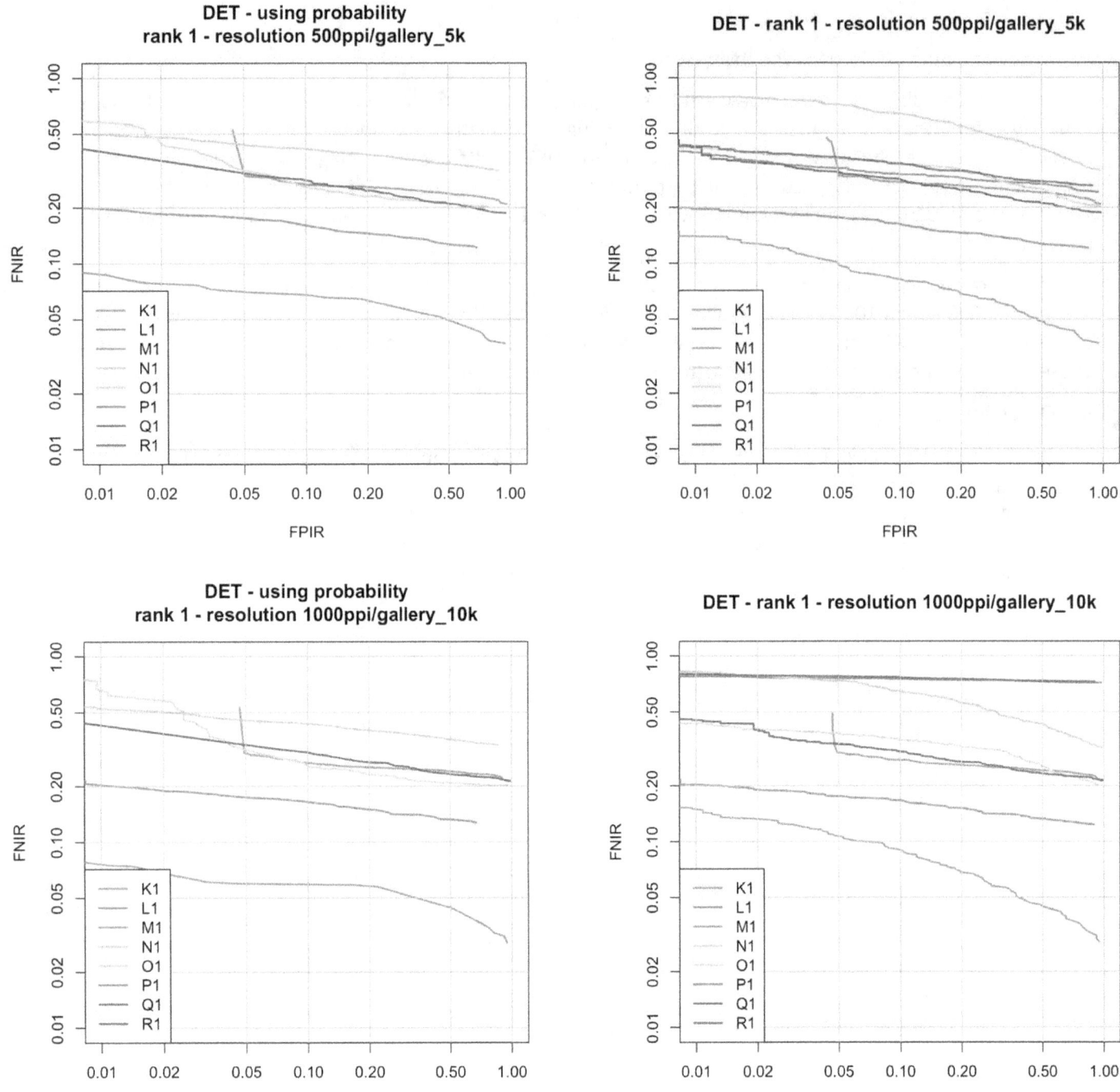

Figure 27: DET at rank 1 of all SDKs comparing between results using probability scores (left) to results using raw matcher scores (right) for two test cases.

5 Fusion

Biometric fusion involves consolidating information from multiple biometric sources. For example, multiple latent impressions of the same person can be combined to improve matching accuracy. Although fusion can be accomplished at various stages during the identification process, this report only addresses fusion at the rank and score level, which occur post-matching. The fusion problem in this context involves consolidating the candidate lists from different searches into a single fused candidate list, preferably with the correct mate at better position on the fused list. Recent work on fusion has been reported for latent fingerprints [12].

Score level fusion uses the raw score associated with each returned candidate to perform fusion, and fusion at the rank level uses only the relative position of the candidate in the unfused candidate list. If the candidate lists were generated by different matching algorithms, as is the case with multi-algorithm fusion, the raw scores must be normalized before they can be directly compared. The rank can be used in place of the raw score to avoid the normalization requirement. However, the rank does not contain as much information as the raw score and may therefore lead to smaller performance gains.

Fusion can also be accomplished at the feature level, which occurs prior to matching. A method of feature-level fusion that is sometimes applied to fingerprints is referred to as image "mosaicking" [13]. The technique involves combining the valid areas of two or more images of the same fingerprint into a new aggregate image. Since the constructed image combines the useful information from each of the individual fingerprint images, it is expected to perform better when matched. Fingerprint examiners sometimes use a similar strategy when they markup a latent image. If an area of a latent image it too poor to identify minutiae, the examiner will look at another latent image of the same fingerprint to extrapolate the location of minutiae within the poor quality area. When mosaicking is performed at the minutia level, it is known as "template consolidation."

Mosaicking is often impractical for the latent identification problem for several reasons. Firstly, multiple latent impressions of the same finger tend to look nearly identical (e.g., when a person flips through a book, he/she tends to leave nearly identical latent impressions on each page). When this occurs, all the latent impressions contain roughly the same information, negating the potential benefit of mosaicking. Secondly, precisely aligning the different images can be difficult if there is little overlapping fingerprint area. An incorrect alignment would misrepresent an individual's finger, leading to a reduced chance of making a correct identification. The safer alternative would be to search each of the impressions separately.

5.1 Multi-Instance Fusion

Multi-instance fusion combines the results of matching two or more latent impressions of a single person. Applying multi-instance fusion to the latent identification problem is very practical since a criminal will often leave several latent impressions at a crime scene. At present, a latent examiner will often run each latent through IAFIS separately. The other latent images are typically used to verify a match that was returned for a particular search. Since AFEM does not require a manual markup of the images, all latent impressions can be searched in parallel with little additional work for the human examiner. For the Phase II dataset, which is somewhat reflective of case work, latent impressions of more than one finger were available for 121 of the 588 subjects. This frequency may be higher than what is typical of actual case work data, since the Phase II set includes only individuals who were identified by IAFIS, and an individual is more likely to be identified if multiple latent impressions are available.

Two latent impressions representing different fingers were selected for each of the 121 subjects. If more than two were available, two were chosen at random. Each latent impression was searched separately and the rank of the correct mate determined. Table 16 shows the frequency at which different combinations of ranks occur for each set of paired latent impressions for two algorithms. The table demonstrates that the correct match is almost always at rank one on at least one of the candidate lists. This behavior occurs for the other algorithms (not shown) as well. This suggests complicated methods of fusion are unnecessary to achieve near optimal results; the fusion method need only place the rank one candidates from the unfused lists at a top position on the fused candidate list. This conclusion may not apply to situations where a longer candidate list is available, or for larger gallery sizes.

	Rank of correct mate for first finger				
Rank of correct mate for second finger	1	2-10	11-20	20-50	Miss
1	111	0	3	1	3
2-10	0	0	0	0	0
11-20	1	0	0	0	0
21-50	1	0	0	0	0
Miss	1	0	0	0	1

(a) M1

	Rank of correct mate for first finger				
Rank of correct mate for second finger	1	2-10	11-20	20-50	Miss
1	86	0	2	1	12
2-10	0	0	0	0	0
11-20	0	0	0	0	0
21-50	1	0	1	0	0
Miss	14	0	2	0	3

(b) K1

Table 16: Two distinct latent impressions were selected for each subject and the rank of the correct mate determined for each latent. The tables show the frequency at which different combinations of ranks occur for each set of paired latent impressions for M1 (a) and K1 (b). Gallery size is 5,000; latent image resolution is 500 ppi.

The Borda count method [14] of fusion was used to generate all fusion plots. Borda count assigns points to each candidate based on rank, with better ranking candidates receiving more points. If a subject appears on more than one list, the points for that subject are summed. The fused candidate list is generated by sorting the subjects by their points such that the subject with the most points is assigned the best rank. Ties are broken randomly to maintain a strict ordering.

Figure 28 shows the result of performing two-finger fusion using the Borda count method for each algorithm. The rank 10 "hit rate" refers to the fraction of searches that placed the correct mate at one of the top 10 positions on the candidate list. A substantial performance improvement occurs for every algorithm. In addition, several algorithms perform almost as well when two fingers are used as M1 using a single finger. The figure may give a false impression of the potential for fusion to improve matching results for two reasons. Firstly, multi-instance fusion can only be performed if two distinct latent impressions are available for a subject, and the Phase II dataset suggests that most of the time only a single latent impression is available. Secondly, although a better rank ordering is achieved, the overall identification rate of case work is unlikely to be affected, since an examiner is still expected to make a positive identification as long as the correct mate is on at least one of the candidate lists.

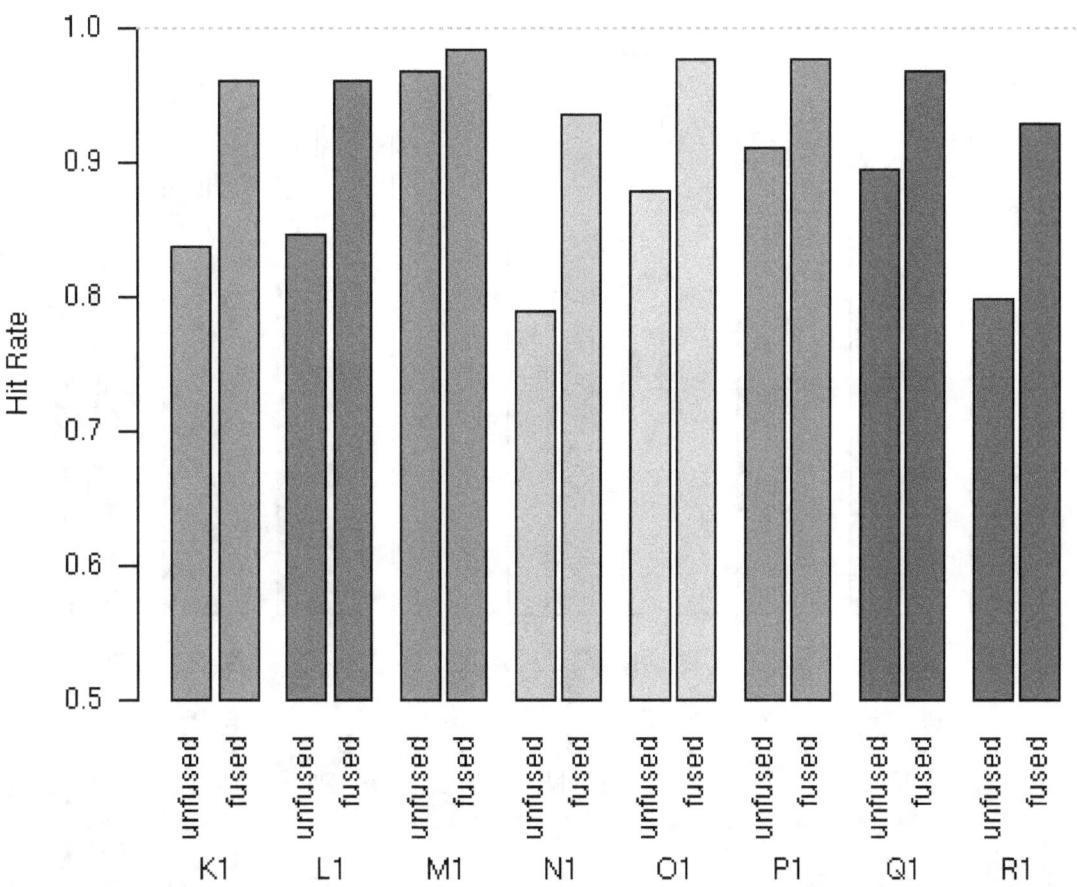

Figure 28: Comparison of rank 10 performance for single finger and two-finger fusion for each algorithm. Gallery size 5,000; latent image resolution 500 ppi.

A possible application of multi-instance fusion is to allow the examiner to send all available latent impressions for a subject to IAFIS. The Phase II dataset contained 467 subjects with only one distinct impression, 121 subjects with two or more distinct impressions, 34 subjects with three or more distinct impressions, and 11 subjects with 4 or more distinct impressions. Figure 29 shows the results of applying Borda count fusion to all of the available latent impressions for each subject. If only one impression was available, no fusion was performed and the unfused candidate list was used. Since only one impression was available for the majority of subjects, the performance improvement is not as pronounced as in Figure 28. Nevertheless, the performance improvement demonstrates that multi-instance fusion can be used to improve rank ordering, thus reducing workload on the human examiner.

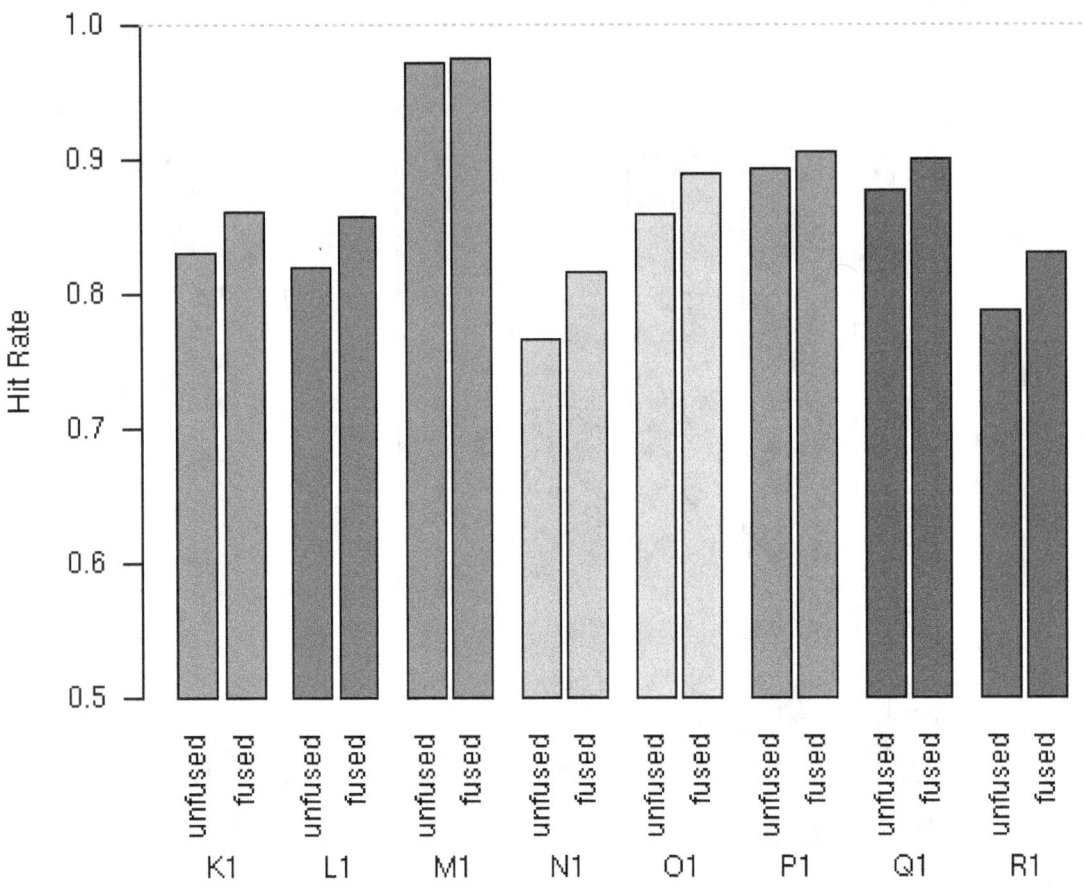

Figure 29: Comparison of rank 10 performance for single finger and for fusion using all available latent impressions per subject. Gallery size 5,000; latent image resolution 500 ppi.

On occasion, two latent impressions may be mistakenly assigned to a single subject when they are, in fact, from different subjects. The examiner would most likely notice such a problem when attempting to verify a "match". When comparing each of the latent impressions to the corresponding ten-print record of the matching subject, the examiner would discover that some of the latent impressions match while others do not. Applying multi-instance fusion to this type of situation is expected to reduce the chances of positively identifying any of the subjects represented by the latent impressions. Figure 30 shows the effect of applying two-finger fusion when the second impression is from a different subject that is not in the gallery set. The figure shows a moderate drop in the hit rate for most of the algorithms. However, the drop is no more than 5 percent for 7 of the 8 algorithms and no more than 10 percent for the remaining algorithm. The figure does not demonstrate what would occur if both subjects were present in the gallery. Nevertheless, the results suggest multi-instance fusion is somewhat robust to situations where two latent impressions are mistakenly assigned to the same subject.

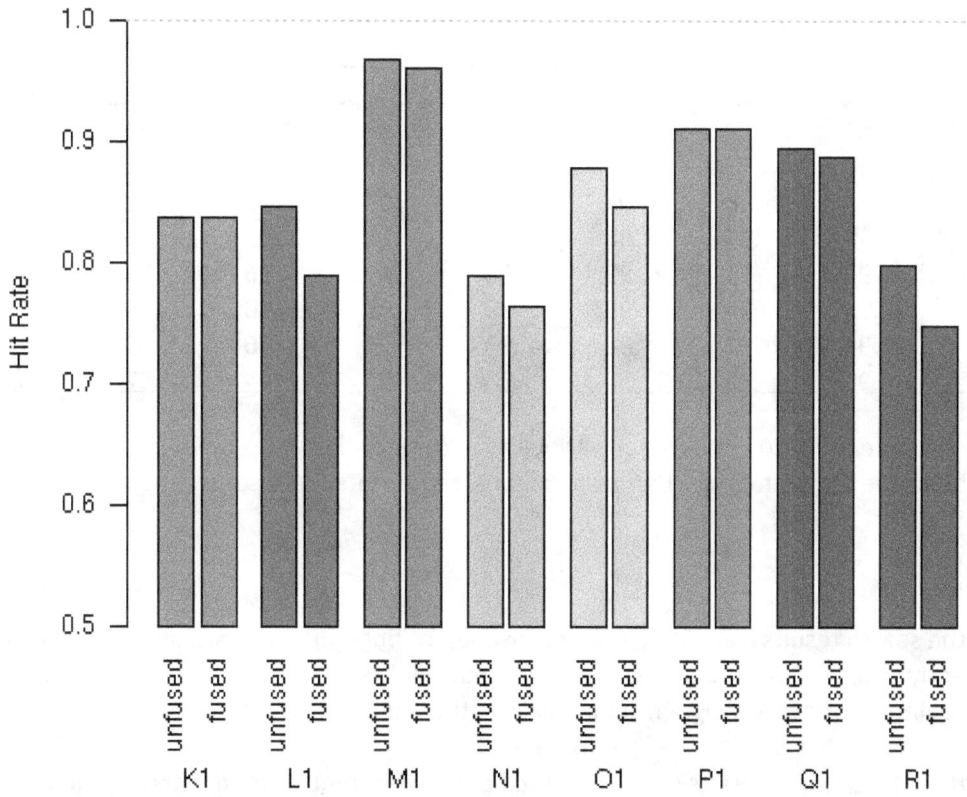

Figure 30: Rank 10 performance for single finger and two-finger fusion when the second latent impression is from a different subject not in the gallery. This might occur if two impressions from different subjects were mistakenly assigned to the same subject.

5.2 Multi-Algorithm Fusion

Multi-algorithm fusion in this context refers to searching the same latent impression using multiple matching algorithms. Unlike multi-instance fusion, it has the advantage of not requiring additional latent information, but it is more computationally intensive. In this section, multi-algorithm fusion is applied to the Phase II data.

Matching performance is highly dependent on the quality of the latent impressions. A difficult latent for one algorithm tends to be difficult for others as well. Thus, a certain amount of correlation between matching algorithms is expected. This is reflected in the candidate lists returned by the different algorithms, which are often very similar with regard to their placement of the correct match. Nevertheless, Table 17 shows that performance can be improved by combining algorithms. In particular, the hit rate can be improved beyond what M1 is able to achieve without fusion. In addition, M1 by itself achieved a hit rate better than any other two algorithms combined (not including M1). However, the statistical significance of these results could not be established.

	K1	L1	M1	N1	O1	P1	Q1	R1
K1	.84	.90	.98	.90	.94	.95	.94	.90
L1	.90	.83	.98	.89	.93	.94	.93	.88
M1	.98	.98	.97	.98	.99	.98	.98	.98
N1	.90	.89	.98	.79	.92	.93	.93	.88
O1	.94	.93	.99	.91	.87	.96	.94	.92
P1	.95	.94	.98	.93	.96	.90	.96	.94
Q1	.94	.93	.98	.93	.94	.96	.88	.92
R1	.90	.88	.98	.88	.92	.94	.92	.81

Table 17: Rank 10 identification rates when various pairs of algorithms are combined by applying Borda count fusion to the candidate lists. The main diagonal represents the cases where no fusion occurs.

Key Observations:

- Fusing the search results (candidate lists) for two latent fingerprints from the same subject consistently improved the identification rate for each of the eight SDKs. This increase in performance can be realized without any significant increase in the human examiner's workload due to AFEM being used.
- Fusing the latent search results (candidate lists) of two different SDKs consistently improved the identification rate. The amount of improvement varied between the systems being fused due to the search results of some systems being more correlated than others.

6 References

[1] S. S. Wood and C. L. Wilson. Studies of Plain-to-Rolled Fingerprint Matching Using the NIST Algorithmic Test Bed (ATB). NISTIR 7112. April 2004. ftp://sequoyah.nist.gov/pub/nist_internal_reports/ir_7112.pdf

[2] NIST Latent Fingerprint Homepage http://fingerprint.nist.gov/latent/

[3] V. Dvornychenko. Concepts of Operations (CONOPS) for Evaluation of Latent Fingerprint Technologies (ELFT), Rev. E, 1 Nov 2007. http://fingerprint.nist.gov/latent/elft07/elft_p2_concept.pdf

[4] V. Dvornychenko. Summary of the Results of Phase I ELFT Testing, 24 Sept. 2007. http://fingerprint.nist.gov/latent/elft07/phase1_aggregate.pdf

[5] P. Grother, M. McCabe, C. Watson, M. Indovina, W. Salamon, P. Flanagan, E. Tabassi, E. Newton, and C. Wilson. Performance and Interoperability of the INCITS 378 Fingerprint Template. NISTIR 7296. March 2006. http://fingerprint.nist.gov/minex04/minex_report.pdf

[6] C. Watson et al. NIST Fingerprint SDK (Software Development Kit) Testing. NISTIR 7119. June 2004. http://fingerprint.nist.gov/SDK/

[7] C. Wilson et al. Fingerprint Vendor Technology Evaluation (FpVTE) 2003. NISTIR 7123. June 2004. http://fpvte.nist.gov/

[8] R. Capelli, D. Maio, D. Maltoni, J. L. Wayman, and A. K. Jain. Performance evaluation of fingerprint verification systems. IEEE Transactions on Pattern Analysis and Machine Intelligence, 28(1):3–18, January 2006.

[9] E. Tabassi, C. L. Wilson, and C. Watson. Fingerprint Image Quality (NFIQ). NISTIR 7151. August 2004. ftp://sequoyah.nist.gov/pub/nist_internal_reports/ir_7151/ir_7151.pdf

[10] E. Tabassi and P. Grother. Recommendations on Biometric Quality Summarization across the Application Domain. NISTIR 7422. May 2007. http://www.itl.nist.gov/iad/894.03/quality/reports/enterprise.pd

[11] Johnson, A., J. Sun, and A. Bobick, "Predicting Large Population Data Cumulative Match Characteristic Performance from Small Population Data," 4th International Conference on Audio and Video Based Biometric Person Authentication (AVBPA 2003), University of Surrey, Guildford, UK June 9-11, 2003.

[12] Jianjiang Feng, Soweon Yoon, and Anil K. Jain, Latent Fingerprint Matching: Fusion of Rolled and Plain Fingerprints, to appear in Proc. International Conference on Biometrics 2009.

[13] A. Jain and A. Ross, Fingerprint mosaicking, in proceedings of IEEE International Conference on Acoustics, Speech, and Signal Processing, 2002. Volume: 4, pp 4064--4067 vol.4

[14] J. Kittler, M. Hatef, R. Duin, and J. Matas. On combining classifiers. IEEE Trans. Pattern Analysis and Machine Intelligence, 20(3), March 1998.

[15] J. Wayman, A. Jain, D. Maltoni, and D. Maio. Biometric Systems. Springer, 2004 - Chapter 6 discusses the modeling of large-scale identification systems.

[16] V. Dvornychenko. Latent fingerprint system performance modeling. Proceedings of the SPIE, Volume 6812, pp. 68120C-68120C-11 (2008).

[17] FBI Electronic Fingerprint Transmission Specification (EFTS), Version 7.1, http://www.fbibiospecs.org/ebts/default.htm

Appendix A– Modelling Effect of Gallery Size

A.1 Scalar Performance Measures

In comparing the performance of different systems it is convenient to be able to do so via a single number (scalar). In this section three such measures are introduced (M_{first}, M_{list} and M_w).

M_{first} is the percentage of searches which resulted in the true match appearing in first (top) place on the candidate list. M_{first} takes on values between 0 and 1 (or 0 to 100%).

The related measure is M_{list}, the percentage of searches which appear (anywhere) on the candidate list. M_{list} also takes on values between 0 and 1 (or 0 to 100%).

M_{first} tends to underestimate performance, while M_{list} provides an overoptimistic value. This is the motivation for introducing the third measure, M_w.

M_w makes allowance for a "hit" in other positions than first place. (These are often referred to as "secondary hits".) However, these secondary hits are weighted less than a hit in first position. The weights are taken to be the reciprocal of the list position. Thus, a hit in first place receives full value ($w = 1$), but one in second position receives only half value ($w = 1/2$); one in third position is given weight ($w = 1/3$), etc. (These weights are also known as "roll-off factors.") As is the case for M_{first} and M_{list}, M_w is always between 0 and 100%. It is easy to see that $M_{first} \leq M_w \leq M_{list}$.

There are two justifications for this weighting/roll-off scheme. The first is that that secondary hits tend to "fall off" (disappear) from the candidate list if the gallery is greatly enlarged. For example, a hit in seventh position has six imposters ahead of it. If the gallery were to be (say) doubled, the expected number of impostors ahead of this hit would be 12. Now the hit could be expected to appear in 13th position. If we suppose the candidate list to be of length ten, the true mate would "fall off" the list.

The second justification for rolling off inversely as the list position takes into account human effort expended. To examine a subject in tenth position (say) requires that the nine candidates ahead of it be examined first. Hence, roughly speaking, the amount of work which must be performed is ten times as large as for a subject in first position. There is therefore justification for considering a subject in tenth position to be worth only 1/10 as much as one in first position.

A number of investigators have suggested different schemes for the roll-off weights. One such proposed method is to roll of inversely as the square-root of the ranking as opposed to inversely as the first power of the ranking [15]. The thinking is that the $1/k$ (k is the ranking) "punishes" lower places too much, and that $1/\sqrt{k}$ is "softer." This may have some merit, but additional experimental support is required. Another proposed scheme is to roll off as $(L+1-k)/L$, where L is the list length. This scheme suffers from the shortcoming that, if L is increased, the value of the metric goes up, even if there were no additional "hits" on the expanded candidate list.

A.2 Power-Law Model

It is desirable to have a performance measure which captures the intrinsic merit of the system, independent of the size and quality of the gallery used in performing measurements. Ideally the performance measure should factor in the difficulty of the searches.

One model which has gained recognition is the power-law. This particular model can be derived/justified theoretically in a number of ways [16]. A particular merit of this model is its simplicity. The probability that the true mate appears on the candidate list is given by:

$$P(L) = (L/N)^\alpha \qquad \ldots (A\text{-}1)$$

Where, L = length of candidate list, N = database size, and α is a parameter characterizing the intrinsic quality of the system and the difficulty of the data. For a "good" system α must be quite small, say $\alpha < 0.1$. Note that $\alpha = 1.0$ represents a random-guessing (zero information) system.

From equation (A-1) we can conclude that the probability of being in first position is $1/N^\alpha$; while the probability of obtaining a hit in position k ($1 < k \leq L$) is

$$p(k) = (k/N)^\alpha - ((k-1)/N)^\alpha \qquad \ldots (A\text{-}2)$$

We have already introduced the weighted metric (M_w). This metric is defined by assigning one point if the mate is in the top position, ½ point if in second position, and so on. This measure provides a finer measure in performance than simply counting the number of mates reported in first place (P(1)). M_w is particularly useful when gauging relatively small changes in performance (for example by doubling the gallery size), as it provides a smoother estimate.

Using eq. (A-2) we can compute the theoretical value of M_w. To a high degree of approximation it is given by

$$M_w = \{1 + (\alpha/(1-\alpha))*(1 - L^{\alpha-1})\} / N^\alpha \qquad \ldots (A\text{-}3)$$

This expression is valid for all values of α in the range 0 to 1.0. For small α and large L (say L>10) the following simplification may be used:

$$M_w = (1 + \alpha)/ N^\alpha = (1 + \alpha)P(1) \qquad \ldots (A\text{-}4)$$

It can be shown that M_w always taken on values between 0 and 1.0, and that $M_w \geq P(1)$. (A result that is obvious from eq. (A-4).)

Note that for high performance systems, M_w and P(1) are nearly the same. (The largest difference between M_w and P(1) occurs when $\alpha = 1/\ln(N)$. Assuming N = 100,000 fingers, this gives $\alpha = .09$. In this event P(1) = 37%, while M_w = 40%. In fact $\alpha = .09$ characterizes a relatively poorly performing system. For small α the two metrics are even more similar.)

A.2.1 Figure of Merit (FOM):

The α of equation (A-1) is a characteristic parameter of a system, in the sense that it captures the intrinsic quality of the system (including database). In particular, α provides a type of performance measure, independent of gallery size.

A figure of merit (FOM) of a system is a measure capturing the intrinsic merit of a system. Of course any given system may be characterized by many FOMs. Convention dictates that a larger FOM indicates a better system. Thus a FOM of 5 should indicate a better system than an FOM of 2.

The parameter α behaves in the opposite manner – smaller values indicate a better system. We can overcome this by defining an FOM as follows:

$$\text{FOM} = -\ln(\alpha) \qquad \ldots (A\text{-}5)$$

This has the required property that the better system provides a higher FOM. Note also that, with using this definition, a random-guessing (zero information) system has an FOM of zero.

A.2.2 Using the Power Law to Analyze Scalability

An objective of the ELFT Phase II tests was to assess the effect of gallery size on the performance of current latent search technologies. We call this the "scalability problem."

That increasing the size of the gallery results in a performance drop (say as measured by the "hits" in first place) is well known. Qualitatively, this decrease in performance can be understood using fairly elementary arguments. However, creating a precise numerical model is much more difficult.

A following simple argument provides insight into the "scalability problem." Suppose that for a gallery of size N the true mate of a given search appeared in fifth position. This means, of course, that the true mate was outscored by four "impostors." If the gallery were now doubled, a reasonable extrapolation is that the number of impostors (outscoring the true mate) would also double, from 4 to 8. The hit would now be expected to appear in ninth position. Of course this is true only "on the average", and in any given trial the hit might appear in eight place, in tenth place, even further away from ninth place.

Execution time imposes limitations on the size of the dataset which can be employed. The Phase II runs required several weeks to execute. Even a modest increase in gallery size requires careful planning so as to execute in a reasonable time; extremely large datasets are prohibitive. NIST's strategy therefore was to supplement direct calculations with modeling and analysis.

We may use the power-law to analyze "scalability", i.e., the effect of an increase in gallery size. Since the power-law is not exact, the results may be of limited accuracy.

The effect on an increase in gallery is readily computed from eq. (A-1). Assuming that the gallery is increase by a factor of k, we substitute kN for N in eq. (A-1) to obtain

$$P(L) = (L/kN)^\alpha = k^\alpha * (L/N)^\alpha \qquad \ldots (A\text{-}6)$$

So the probability is decreased by a factor of $k^{-\alpha}$. The net decrease in hit rate will be

$$\text{Delta\_P}(L) = (k^{-\alpha} - 1)*(L/N)^\alpha \qquad \ldots (A\text{-}7)$$

For the present case, representative values are: L=1 if only first-place hits are counted; L = 50 if all hits on the candidate list are counted; N = 50K fingers (i.e., 5000 subjects) for the smaller gallery; and N = 100K fingers (i.e., 10,000 subjects) for the larger gallery.

The value of α is obtained empirically by fitting eq. (A-1) to the Phase II test results, and differs for each SDK. The computed values also depend somewhat on the value of L selected (e.g., L=1 or L=50), as well as the value of N (50K or 100K). We compensate for this small variation by averaging over these cases (this is roughly equivalent to a least-squares solution).

The mean value over the SDKs so obtained is α = .0172. (SDKs L1 and R1 are not included in this average due to their curious performance on 1000 ppi latents).

If we now assume L =1, N = 50K, and use the above value of α, we can compute P(1) = 0.826, or 82.6%. This is the correct value for aggregate of first place hits for the six best SDKs.

Suppose now we would like to estimate the effect of doubling the repository. Using the same parameters of before, but doubling N to 100K, we compute P(1) = 0.816, or 81.6%. This suggests that approximately 1% of hits can be expected to be lost from first position.

A.2.3 Elementary Error Analysis

As we have seen, the actual decrease in performance due to doubling the gallery is quite small. Unfortunately this has an adverse effect on the statistics.

For example, assume that the number of searches having mates is 813. We then expect approximately 672 hit in first place; doubling the gallery will lose about 8 hits from top position (more precisely, 8.4 on the average).

A simple calculation then shows the standard error of the number of hits lost is about 2.9, or about 35% of the hits lost from top position. The conclusion is that the extrapolation uncertainty is quite large.

A.2.4 Test Results

The following table shows the values of α computed for the eight SDKs (L1 and R1 are included). In the table, α is referred to as "alpha_bar" as it represents the average of several values.

SDK	alpha_bar
K1	0.020806
L1	0.136254
M1	0.002054
N1	0.031001
O1	0.018518
P1	0.013088
Q1	0.017759
R1	0.135665

Table A-1: Computed (estimated) values of alpha (α) for the eight SDKs

The graph below (Figure A-1) provides a comparison between the observed and predicted decrease in performance due to doubling the size of the gallery. The values shown represent P(1;50K) − P(1;100K), and are given as a percentage. The bars labeled "observed" are the actual test results, while "computed" are based on calculations using the method outline above. As mentioned, in interpreting these results it should be kept in mind that the statistical fluctuations (i.e., uncertainty) are quite large.

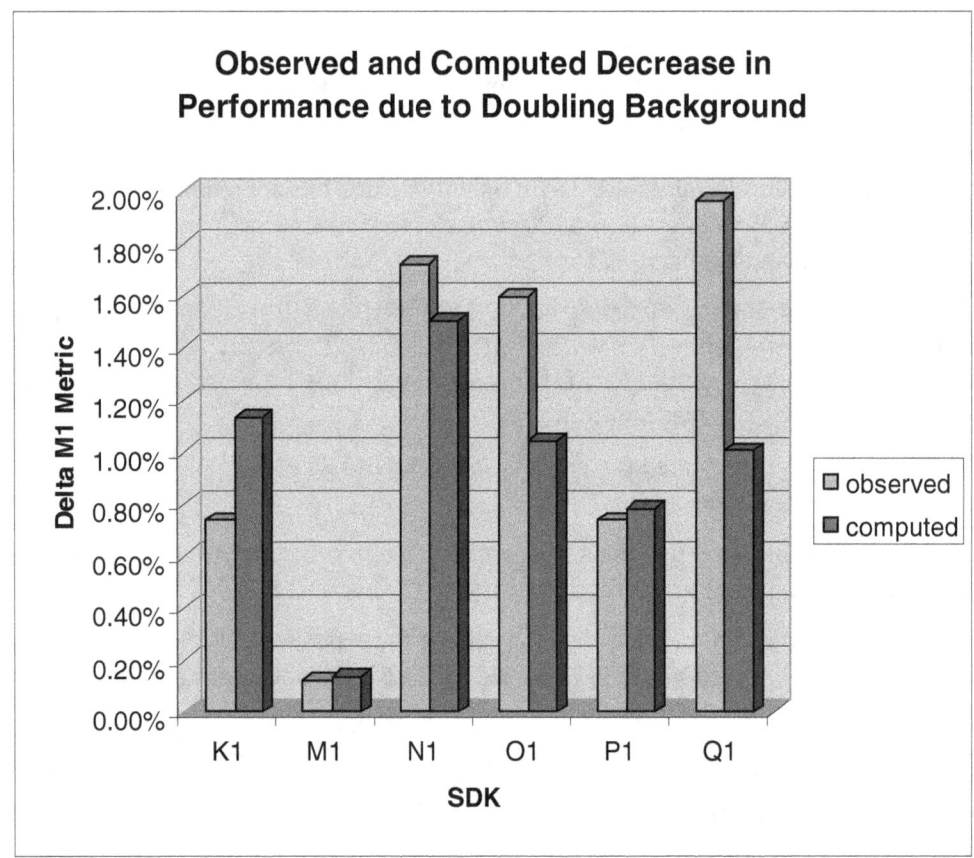

Figure A-1: Observed and computed scaling results

Note that in general the agreement is good, except for Q1; even for Q1 the predicted result is of the right order of magnitude.

In the previous analysis we looked at the effect of increasing in the gallery by a factor of two. This modest increase in the ELFT Phase II testing was dictated by software execution time limitations. For a convenient "rule of thumb" it is better to use a factor of ten when testing for the effect of scaling gallery size. A computed estimate of this parameter is shown in the following table in the column labeled "delta_10". (once again, SDKs L1 and R1 are excluded.) The table also shows the computed α, as well as a derived Figure of Merit (FOM). Larger FOMs indicate better performance.

	alpha_bar	delta_10	FOM2
K1	0.02081	4.68%	3.87
M1	0.00205	0.47%	6.19
N1	0.03100	6.89%	3.47
O1	0.01852	4.17%	3.99
P1	0.01309	2.97%	4.34
Q1	0.01776	4.01%	4.03
Average	**0.01720**	**3.86%**	**4.32**

Table A-2: Predicted decrease in performance due to factor of ten Increase in gallery size

Based on the above table, Figure A-2 shows the projected performance for large databases. Rather than showing a separate graph for each SDK, we present the performance for a hypothetical "averaged" matcher (whose performance is the "aggregate" or mean of the six). Note that the size of the gallery is given in subjects, rather than fingers. With a background of 1 million subjects, the predicted performance is 75.04%; with 5 million subjects, the predicted performance is 72.94%; and with 10 million subjects, the predicted performance is 72.04%.

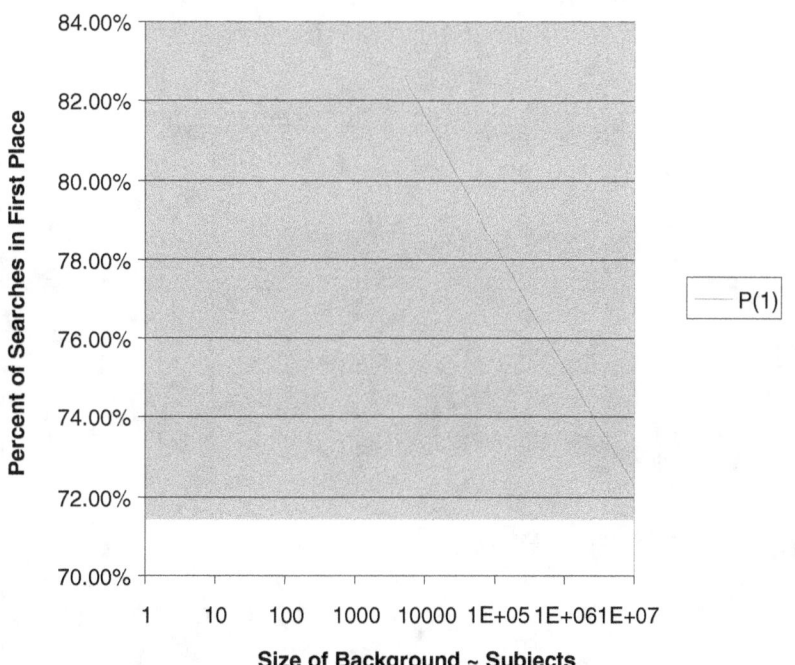

Figure A-2: Predicted performance of hypothetical "average" matcher as a function of gallery size

Key Observations:

- The power law appears to be a reasonable "order of magnitude" predictor of the change in performance due to scaling the gallery.

- Based on calculations, current SDK may perform well even if the gallery is increased by several orders of magnitude.

- The reader is strongly encouraged to not put too much weight in these results. In this case a very simplistic model was used based on the average performance of the eight SDKs in Phase II. The ability of the model to predict the change in performance between the 5K and 10K gallery is promising, but there is no evidence that the model's estimates should be trusted further out. More work in this area is warranted

Appendix B– ELFT Phase II Protocol Description

ELFT is a technology evaluation, similar to [5] and [6], which uses a software-only approach that attempts to minimize the "entry cost" for potential participants regarding implementation of their algorithms. ELFT employs a PC-based approach in which participants may submit either Windows or Linux executable libraries for use with off-the-shelf PC hardware. This approach itself however places limitations on the testing environment (e.g. processor speed, I/O speed, and memory), which may not otherwise exist in operational systems. In response, the ELFT testing protocol seeks to manage these limitations within the goals of the evaluation.

B.1 API

The use of a common testing API promotes efficient and uniform testing and minimizes testing errors. The ELFT Phase II API specification is included in Appendix C, and it defines a set of requirements that all SDKs submitted for ELFT testing must follow. The Phase II API consists of both optional and mandatory elements, and includes:

- Test image format (latent and ten-print)
- Procedural interfaces (function prototypes & data structures)
- SDK library binary format & platform requirements
- Execution speed requirements (in the form of time limits for enrollment and search)

The API used for Phase II was an evolution of the ELFT Phase I API, with the following modifications:

- Revised method for specifying estimated latent print orientation
- Revised error handling and reporting
- Revised specification of image size ranges
- Revised specification of file pathname syntax
- Restriction to non-adaptive matching
- Additional threading support and documentation requirements
- Revised mean execution time limits for enrolling latent and gallery images

B.2 Computer Hardware Platform

Only NIST hardware was used for ELFT testing, and only NIST had access to this hardware. A total of forty-eight Dell 1855 Blades were used for execution of submitted SDKs. The hardware configuration of these machines is as follows:

- Processor:
    - Dual 2.8 GHz/1MB Cache, Xeon (dual-core)
    - 800 MHz Front Side Bus for PE 1855
- Memory:
    - 2GB DDR2 400 MHz (2x1GB) single ranked DIMMS
- Secondary storage:
    - DUAL 73GB 10K RPM, Ultra 320 80 pin SCSI Hard drives (hot plug)

SDKs were submitted in encrypted form. Submitters could specify whether they wished to execute in a Window or Linux environment. Six of the tested SDKs were executable on Windows, and two on Linux. Windows executables were tested on blades running Windows 2003 Server (service pack SR2). Linux executables were tested on blades running Red Hat Enterprise Linux ES Release 4 (Nahant).

Each SDK had access to all of the machine's resources (less an allowance for normal operating system overhead), and at no time during execution did more than one SDK run on the same physical machine. Additionally, all I/O performed to retrieve test images and write output data was performed using locally attached non-shared storage.

Note that the above hardware may not be representative of operational systems. Therefore, all timing requirements and all timing results may not be directly comparable to operational scenarios, which may involve different hardware and software.

B.3   Execution Time Limits

Our procedure for establishing test execution time limits was as follows. Prior to Phase I we requested execution time estimates from all the technology providers. Based on this feedback and observed processing times from past 1:1 tests conducted at NIST, we set the Phase I execution time limits. Using the observed Phase I timing results we then adjusted the limits slightly for Phase II. The execution times chosen are summarized in the following table.

Function	Phase I time limit	Phase II time limit
Enroll ten-print	150 seconds per ten-print (10 sec/finger)	150 seconds per ten-print (10 sec/finger)
Enroll latent image	350 seconds/image	600 seconds/image
Search compare	0.2 seconds per ten-print	0.25 seconds per ten-print

Table B- 1: Phase II SDK execution time limits

Some in the biometrics community have expressed an opinion that the allowable execution time is too generous. Their principal argument is that the allowable time far exceeds that which would be available in operational scenarios. NIST's position is that at this stage of testing we are not attempting to simulate operational scenarios. Rather, we are trying to establish best possible performance within a "reasonable" period of time for testing. In due course we will address operational scenarios.

B.4   Test Execution

The Phase II test execution protocol will be briefly discussed in this section. More details are available in the CONOPS [2].

Executable modules for testing were constructed from two sources: 1) technology provider-supplied software in the form of a Software Development Kit (SDK), and 2) NIST-supplied software. The submitted SDKs were executable in either a Windows or Linux environment, based upon the technology provider's preference. The core of the executable module is of course derived from the SDK. The part supplied by NIST is mainly concerned with the image retrieval and manipulation. The executable module itself is known as the NIST Test Application.

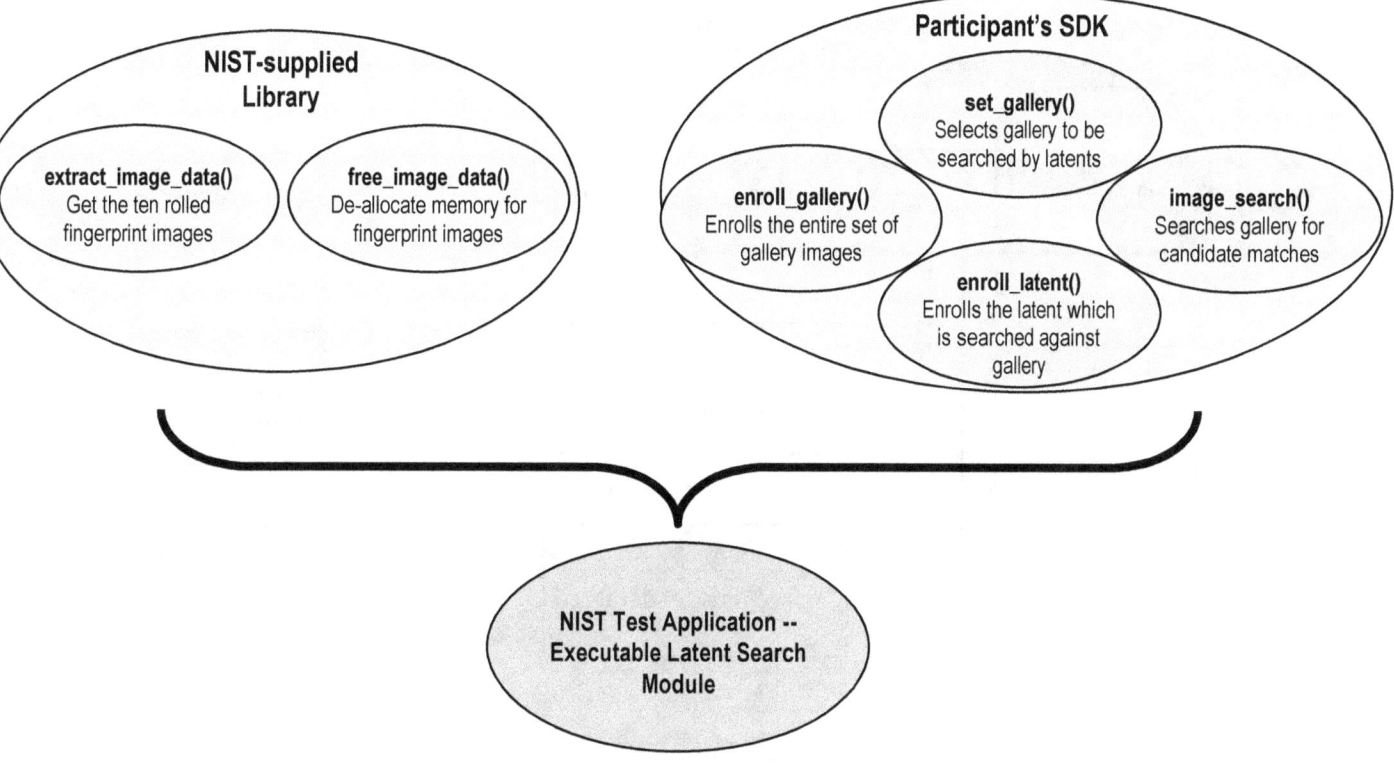

Figure B-1: Executable software is a combination of technology provider-supplied (SDK) and NIST-supplied software

Once constructed (linked), the executable modules (one per SDK) are executed in three stages, each being a distinct operational mode of the NIST Test Application

1. Gallery enrollment
2. Latent feature extraction
3. Latent search against gallery using extracted features

Stage 3 depends on data output from the successful completion of stages 1 and 2; however stages 1 and 2 are independent and may be executed concurrently on separate processors.

The first stage performs "enrollment" of the gallery fingerprints. This process converts images into proprietary "feature" representations. The output of this enrollment stage is at the discretion of the technology provider, except that all extracted data must be written into a single directory specified by the NIST Test Application A gallery consists of a series of ten-prints (i.e. a record with ten fingerprints per subject), with each stored in a file consisting of ten Electronic Fingerprint Transmission Specification (EFTS) [17] type-14 records. Each type-14 record contains a single WSQ-compressed fingerprint image. During enrollment, a list of ten-print filenames is input by the NIST Test Application to the SDK. To promote uniform parsing and decompression of these images, NIST supplied "extraction routines" (with interfaces specified in the ELFT Phase II API) for retrieving the individual images.

The following diagram summarizes the gallery feature extraction pass. The calls to NIST-provided image extraction routines are highlighted in blue. All SDKs were executed (in stage 1 processing) so as to enroll four separate galleries, as shown in Table B-2.

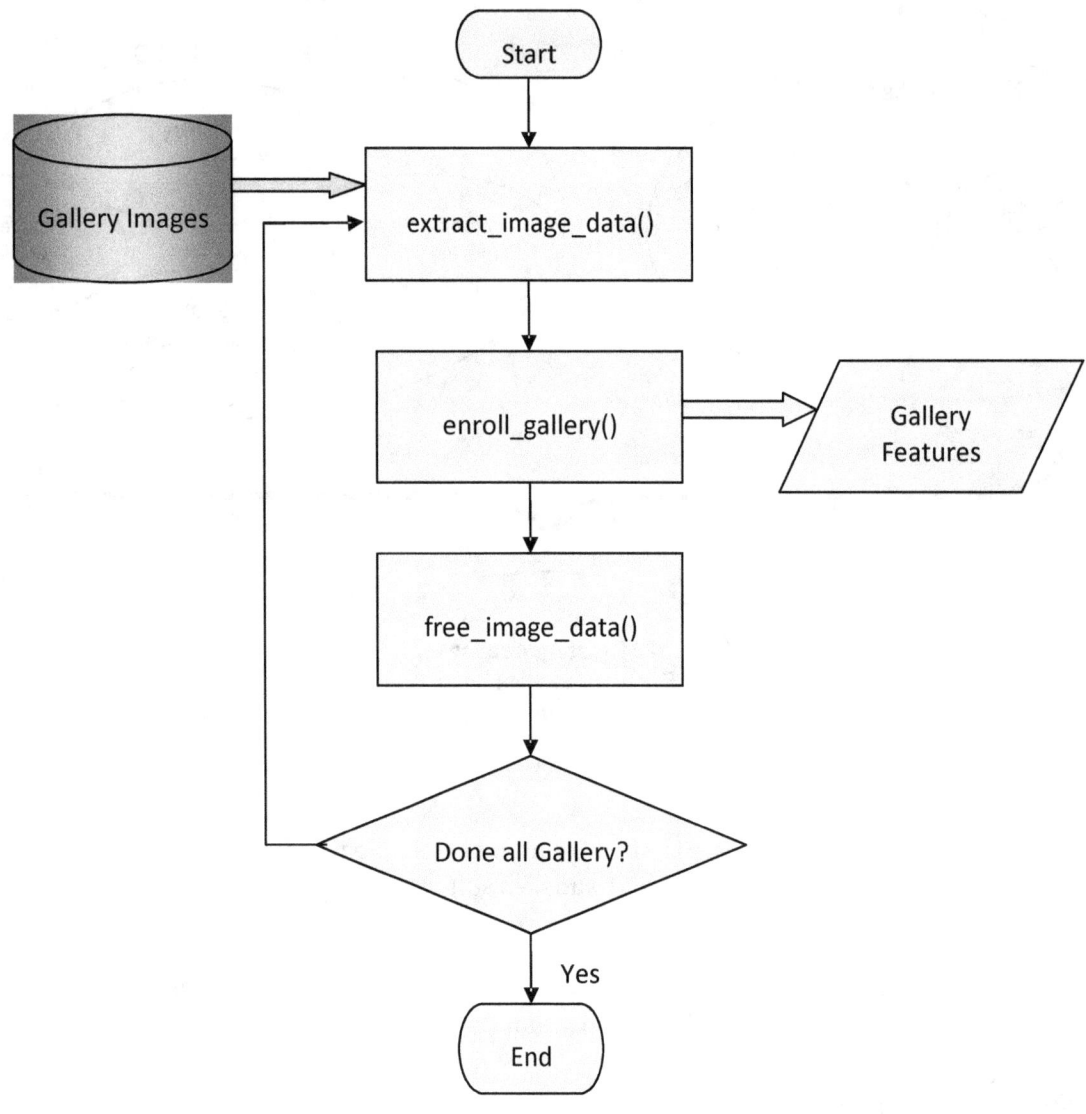

Figure B-2: Stage 1 - Logic and data flow for Gallery enrollment

Gallery Name	Total ten-print records	Non-mate ten-print records (background)	Mate ten-print records (foreground)
G1A	10000	9392	608
G1B	10000	10000	0
G2A	5000	4392	608
G2B	5000	5000	0

Table B-2: All Phase II test Galleries (enrolled in stage 1)

The second stage of execution generates a set of proprietary features for all latent fingerprint images. The specific features extracted and their formats are at the discretion of the technology provider, and may even include the original latent fingerprint image itself in its entirety. The latent fingerprint images and Region of Interest (ROI) masks input to the SDK by the NIST Test Application are in uncompressed "raw" format. The details of the latent enrollment process also depend on whether a latent ROI mask is specified or not, as shown by the following figure.

All SDKs were executed in stage 2 so as to extract features from two sets of 835 distinct latent fingerprint images. The first set, L1, consists of images at 500 ppi resolution. The second, L2, consists of images at 1000 ppi resolution. A third set, labeled L3, consists of the extracted features from the 1000 ppi latent images accompanied by an ROI mask.

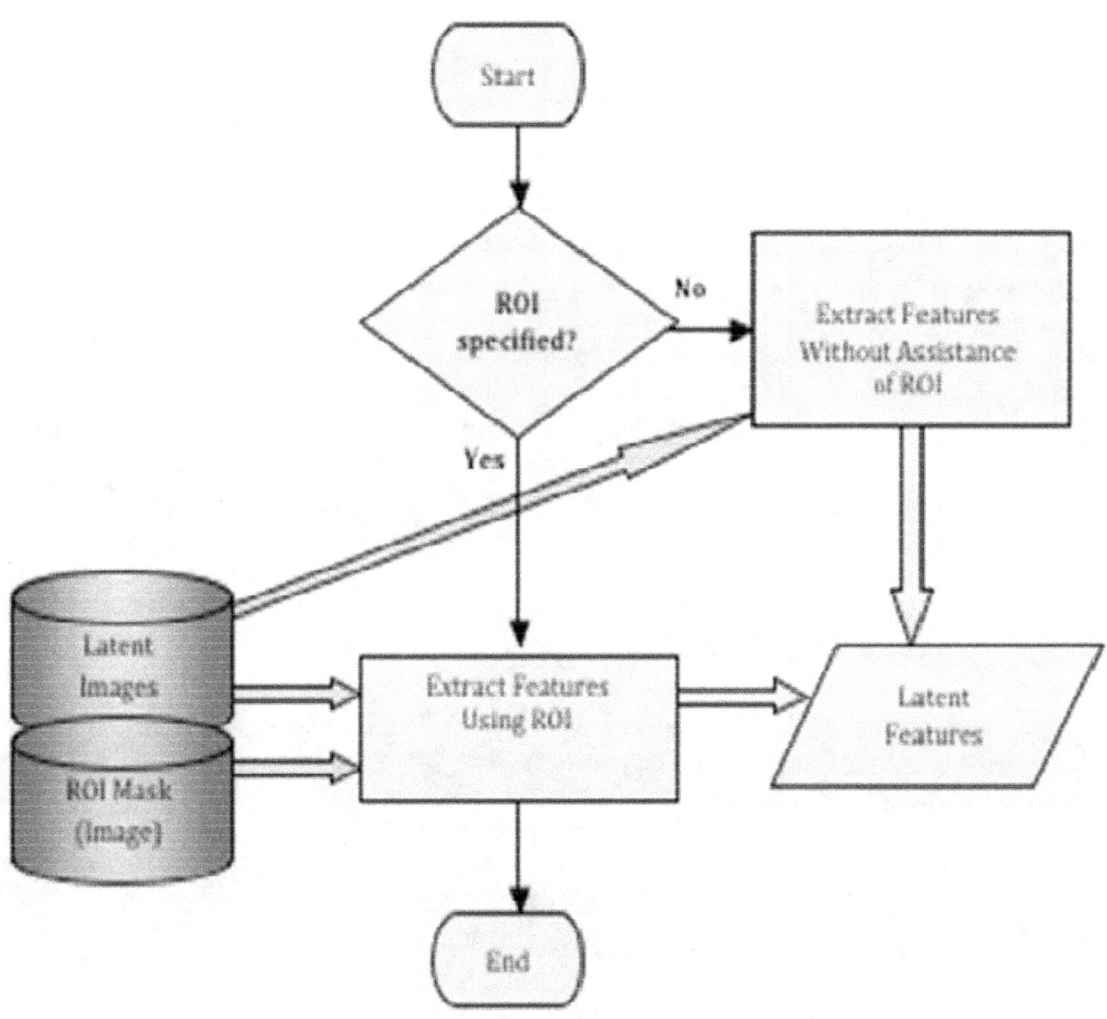

Figure B-3: Stage 2 - Logic and data flow for latent feature extraction

In Stage 3 processing, each latent's proprietary feature set (created in stage 2) is fetched and searched against the enrolled gallery data (created in stage 1), and a candidate list is returned. The details of the searching process are unregulated by the API. In particular, technology providers may invoke multi-stage algorithms within the top-level search function called by the NIST Test Application The following diagram summarizes the logic and data flow during the matching pass.

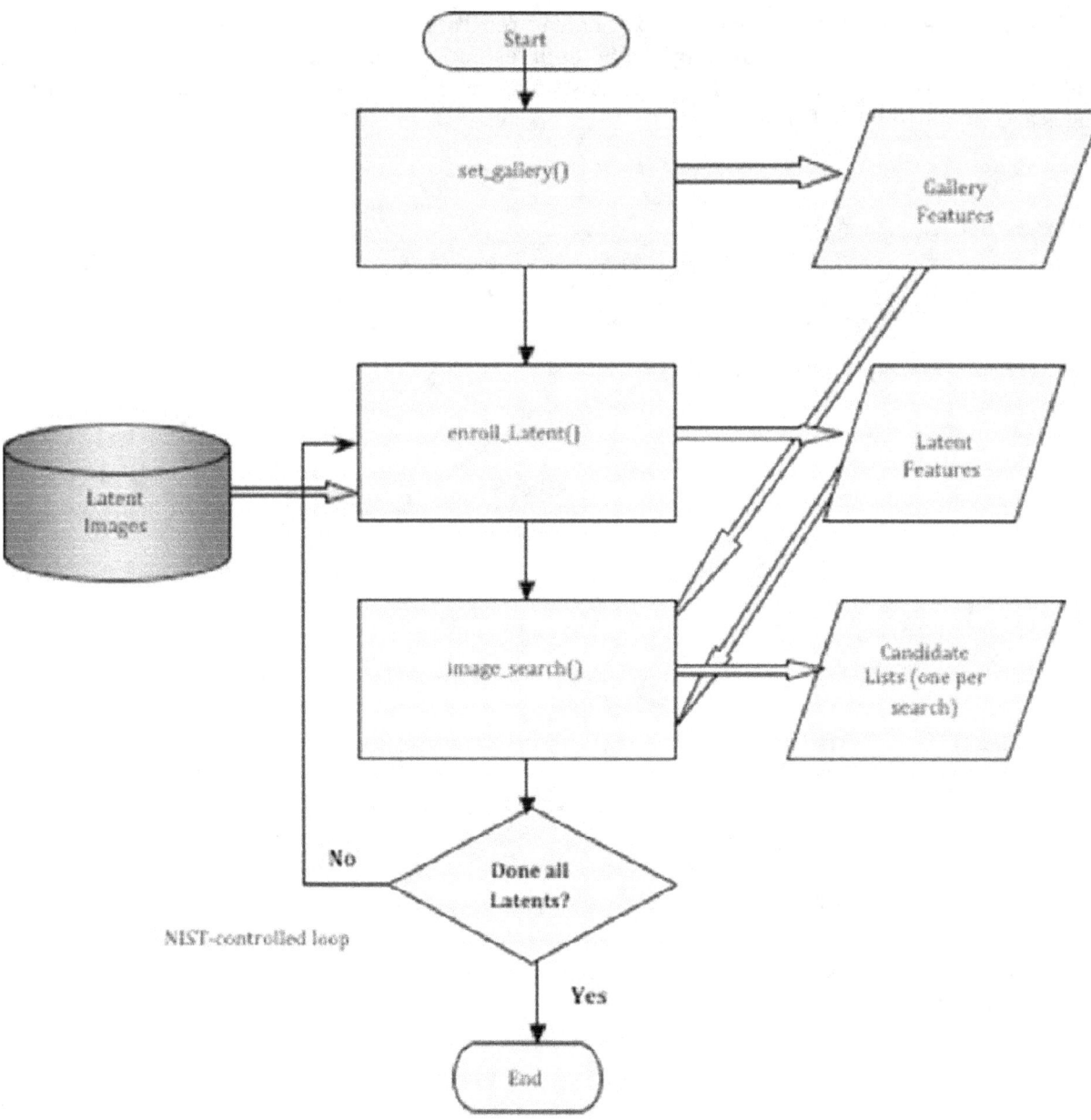

Figure B- 4: Stage 3 - Logic and data flow for Latent image search of Gallery using extracted features

Each SDK was executed (in stage 3) to perform searches using eight distinct combinations ("search configurations") of enrolled latent and gallery prints as summarized in the table below.

Latent resolution (ppi)	Gallery size (total ten-prints)	Total # of non-mate ten-prints (background)	Total # of mate ten-prints (foreground)	Total # of Latents	Gallery contains mates?	ROI provided?	Search configuration name
500	5000	4392	608	835	Yes	No	L1_vs_G2A
500	5000	5000	0	835	No	No	L1_vs_G2B
1000	10000	9392	608	835	Yes	No	L2_vs_G1A
1000	10000	10000	0	835	No	No	L2_vs_G1B
1000	5000	4392	608	835	Yes	No	L2_vs_G2A
1000	5000	5000	0	835	No	No	L2_vs_G2B
1000	5000	4392	608	835	Yes	Yes	L3_vs_G2A
1000	5000	5000	0	835	No	Yes	L3_vs_G2B

Table B-3: Latent vs. Gallery search configurations

B.5 SDK Conformance Testing

Technology providers in Phase II were permitted to submit a single SDK. Once received by NIST, each SDK was tested for conformance to the ELFT II API by means of a sample "validation set" of latent and gallery ten-print images. Latent and gallery images were input to the SDKs by NIST with the objective to identically reproduce the candidate lists submitted by the technology providers. All candidate lists were required to be reproduced identically to ensure proper installation and expected operation of the technology provider's software before the SDKs were accepted by NIST for testing. In several cases, due to problems encountered (non-conformance, crashes, etc.), several iterations of SDK and candidate lists were solicited from a given technology provider before SDK acceptance. All technology providers in ELFT Phase II eventually passed conformance testing of their submitted SDKs.

Appendix C– ELFT Phase II API Specification

Introduction

The Latent Fingerprint SDK Test provides a means of determining core search performance of latent-fingerprint matchers. This document specifies all SDK interfaces and functionality as well as the data formats used for this test.

There will be minimal human involvement during the actual execution of the test. A small amount of human assistance will probably be required to prepare the data. All such assistance will be provided indirectly by NIST, and may include:

a) Crop and orient certain latents.

b) Provide a region-of-interest.

c) Provide latent experts for examining potential consolidations.

Those wishing to submit software for Latent Fingerprint SDK testing shall be required to provide NIST with an SDK (Software Development Kit) library which complies with the API (Application Programmer Interface) specified in this document.

C.1 Fingerprint Image Data

C.1.1 Format

The SDK must be capable of processing fingerprint images supplied to the SDK in uncompressed raw 8-bit (one byte per pixel) grayscale format. The image data shall appear to be the result of a scanning of a conventional inked impression of a fingerprint. Figure C-1 illustrates the recording order for the scanned image. The origin is the upper left corner of the image. The x-coordinate (horizontal) position shall increase positively from the origin to the right side of the image. The y-coordinate (vertical) position shall increase positively from the origin to the bottom of the image.

Scan Representation

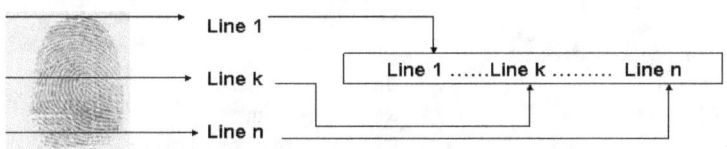

Figure C-1: Order of scanned lines

Raw 8-bit grayscale images are canonically encoded. The minimum value that will be assigned to a "black" pixel is zero. The maximum value that will be assigned to a "white" pixel is 255. Intermediate gray levels will have assigned values of 1-254. The pixels are stored left to right, top to bottom, with one 8-bit byte per pixel. The number of bytes in an image is equal to its height multiplied by its width as measured in pixels; there is no header. The image height and width in pixels will be supplied to the SDK as supplemental information.

C.1.2 Resolution, Dimensions and Orientation

The latent fingerprint images will employ either 500 or 1000 ppi resolution (both horizontal and vertical). All background fingerprint images will employ 500 ppi resolution (both horizontal and vertical). The precise resolution for each individual image will be specified to the SDK via the API.

All fingerprint images used for the background will vary from 150 to 1000 pixels in both width and height dimensions. All latent images at 500 ppi will vary from 150 to 1000 pixels in both width and height. All latent images at 1000 ppi will vary from 150 to 2000 pixels in both width and height. The precise dimensions of each individual image will be specified to the SDK via the API.

All latent fingerprint images used for Phase II testing may vary in orientation over the full angular range (0 to 359). The estimated orientation and range of uncertainty of individual latent prints may be specified to the SDK via the API. Otherwise, the orientation is specified as "upright" +-180 degrees. No information will be specified to the SDK regarding the orientation of background fingerprint images.

No information regarding the distribution of fingerprint image resolution, dimensions, or orientation within the Phase II dataset is provided in this document.

C.2   Test Interface Description

Participants shall submit an SDK which provides the interfaces defined in section C.2.3. Section C.2.2 defines the interfaces to functions provided by NIST for use by the SDK. Sections C.2.1 and C.2.4 specify the declaration of constants, error codes, data-types and functions used by both.

C.2.1 Declarations

The following are declarations of data types and functions used in the Latent Fingerprint SDK testing interface:

```
////////////////////////////////////////////////////
// Declarations of constants                      //
////////////////////////////////////////////////////

// Impression type codes
#define IMPTYPE_LP      0       // Live-scan plain
#define IMPTYPE_LR      1       // Live-scan rolled
#define IMPTYPE_NP      2       // Nonlive-scan plain
#define IMPTYPE_NR      3       // Nonlive-scan rolled

// Finger position codes
#define FINGPOS_UK      0       // Unknown finger
#define FINGPOS_RT      1       // Right thumb
#define FINGPOS_RI      2       // Right index finger
#define FINGPOS_RM      3       // Right middle finger
#define FINGPOS_RR      4       // Right ring finger
#define FINGPOS_RL      5       // Right little finger
#define FINGPOS_LT      6       // Left thumb
#define FINGPOS_LI      7       // Left index finger
#define FINGPOS_LM      8       // Left middle finger
#define FINGPOS_LR      9       // Left ring finger
#define FINGPOS_LL      10      // Left little finger

////////////////////////////////////////////////////////////
// Declarations for the NIST provided library functions  //
////////////////////////////////////////////////////////////

// Structure to hold a single fingerprint record (image+metadata)
```

```c
struct finger_record
{
        BYTE    impression_type;
        UINT16  resolution;        // Image resolution in pixels/cm
        BYTE    finger_position;
        UINT16  height;            // Image height in pixels
        UINT16  width;             // Image width in pixels
        BYTE    *image_data;       // 8-bit grayscale image data
};
typedef struct finger_record    FINGER_REC;
```

// Extracts 10 fingerprint records from a ten-print (AN2K) file
INT32 extract_image_data(const char *tenprint_filename,
        FINGER_REC **finger_recs);

// De-allocates the memory holding 10 fingerprint records
void free_image_data(FINGER_REC *finger_recs);

//
// Declarations for the SDK provided library functions   //
//

```c
// Structure to hold zero or more candidates returned in a search
struct candidate {
        UINT32 background_index;
        BYTE   finger_position;
        DOUBLE similarity_score;
        BYTE   probability;
        UINT16 num_matching_minutiae;
        BYTE   candidate_quality;
}
typedef struct candidate CANDIDATE;

// Structure to hold list of candidates returned by SDK
struct candidate_list
{
        UINT32     num_entries;
        UINT16     num_latent_minutiae;
        BYTE       latent_quality;
        CANDIDATE  *list;
};
typedef struct candidate_list  CANDIDATE_LIST;
```

// Enrolls the entire set of background images
INT32 enroll_background(const INT32 num_recs,
        const char **filenames, const char *enrollment_dir,
        char *error_msg);

// Selects the current background for latent image searching
INT32 set_background(const char *enrollment_dir);

// Enrolls the latent image
INT32 enroll_latent(const FINGER_REC *latent_finger,
        const BYTE *roi_mask, const UINT16 orientation,
        const BYTE offset, BYTE *enrolled_latent,
        INT32 *enroll_length);

```
// Searches for the latent image in the background
INT32 image_search(const BYTE *enrolled_latent,
        CANDIDATE_LIST *candidates, char *error_msg);
```

C.2.2 NIST Provided Functions

C.2.2.1 Extract Image Data

```
INT32
extract_image_data(const char   *tenprint_filename,
                   FINGER_REC   **finger_recs);
```

Description

This function extracts ten fingerprint image records from a single (AN2K formatted) ten-print record file. The caller shall pass tenprint_filename as a pointer to the fully qualified pathname of an AN2K formatted ten-print record file, and finger_recs as the address of a pointer of type `FINGER_REC` (see 2.1 above).

Upon return finger_recs will contain a pointer to an array of ten `FINGER_REC` structures ordered by finger position from 1 (right thumb) to 10 (left little finger). For any fingers that are missing from the original ten-print record file, the image_data field in the respective `FINGER_REC` will be a NULL pointer.

Example
```
// Example of processing a ten-print record
FINGER_REC *finger_recs;
INT32 status=extract_image_data("image00205.an2k", &finger_recs);
if (status == 0) {
        for (i=0;i<10;i++) {
                if (finger_recs[i].image_data != NULL)
                        process_valid_finger(finger_recs[i]);
                else
                        process_missing_finger(finger_recs[i]);
        }
        free_image_data(finger_recs); // see 2.2.2 below
}
```

Parameters

tenprint_filename (**input**): A pointer to a ten-print record filename.
finger_recs (**output**): The address of a FINGER_REC pointer.

Return Value

This function returns zero on success or a documented non-zero error code otherwise.

C.2.2.2 Free Image Data

```
void
free_image_data(FINGER_REC *finger_recs);
```

Description

De-allocates all memory used by the array of `FINGER_REC` structures specified by finger_recs which was allocated during a call to `extract_image_data()`.

Parameters

finger_recs (**input**): A pointer to an array of FINGER_REC structures.

Return Value

None.

C.2.3 SDK Provided Functions

C.2.3.1 Enroll Background

```
INT32
enroll_background(const INT32  num_recs,
                  const char   **filenames,
                  const char   *enrollment_dir,
                  char         *error_msg);
```

Description

This function performs the conversion of all background ten-print records into a proprietary dataset. No format is prescribed for this data, but it could be a set of proprietary templates. Pre-computation of background data avoids reprocessing of the original images upon subsequent calls to `image_search()`.

The SDK shall use the function `extract_image_data()` (see 2.2.1 above) provided by NIST to extract the raw grayscale image and metadata from each ten-print record file specified in the filenames array. Note that each call to `extract_image_data()` allocates memory to hold the extracted image and metadata, so this memory should be de-allocated using the NIST provided `free_image_data()` (see 2.2.2 above) function when no longer needed.

The format of the filenames pointed to by the filenames array will be canonical Unix style pathnames using forward slash directory separators (e.g. "/mnt1/xyz/foo-22/image00205.an2k").

All data produced by the SDK shall be stored exclusively to the directory specified by enrollment_dir. The contents of this directory are at the discretion of the vendor.

Non-fatal error conditions shall be tolerated and shall ~~not~~ result in pre-mature halting (i.e. non-completion of background enrollment). These error conditions include missing fingers in ten-print records, and failure-to-enroll (FTE) any portion of a ten-print record. If any of the above non-fatal error conditions are encountered, the SDK may optionally return a documented non-zero warning code (after completing background enrollment), though this is not required.

Upon entry the error_msg parameter will point to a pre-allocated and pre-zeroized string buffer of length 513 bytes (512 + 1 for the NULL terminator) which the SDK may use for outputting detailed information regarding fatal errors which have occurred (signaled by a non-zero return code). This may be useful for debugging any problems that might occur after the SDK is received by NIST. For example, if the enrollment process encounters a fatal or non-fatal error during processing of a specific background ten-print record file, the SDK could output an error message including the ten-print record filename to error_msg and return a documented non-zero error or warning code respectively.

Note 1: The order of the ten-print record file names in filenames defines (implicitly) the indexing scheme that shall be used henceforth for recording the ten-print record indices of all candidates returned by **image_search()**. The index of the first ten-print record is 1.

Note 2: During subsequent calls to **image_search()** the SDK is permitted to access the original background images. To support this access, the path information supplied by filenames regarding the original background images should be stored in the proprietary background set in enrollment_dir.

Parameters

num_recs (**input**): The number of ten-print records to enroll.

filenames (**input**): Array of pointers to all ten-print record filenames.

enrollment_dir (**input**): The directory used to store enrollment data output.

error_msg (**output**): Pointer to a detailed error message string.

Return Value

This function returns zero on success or a documented non-zero error code otherwise.

C.2.3.2 Set Background

```
INT32
set_background(const char  *enrollment_dir);
```

Description

This function selects the background that shall be used by all subsequent calls to `image_search()`. The directory specified by enrollment_dir shall contain the enrollment data produced by a prior call to `enroll_background()`.

Parameters

enrollment_dir (**input**): The directory to be used by image_search().

Return Value

This function returns zero on success or a documented non-zero error code otherwise.

C.2.3.3 Enroll Latent

```
INT32
enroll_latent(const FINGER_REC *latent_finger,
        const BYTE      *roi_mask,
              const UINT16    orientation,
              const BYTE      offset,
              BYTE          *enrolled_latent,
              INT32         *enroll_length);
```

Description

This function enrolls the latent image pointed to by **latent_finger**, and writes the enrollment data to the memory location pointed by **enrolled_latent**. The latent image itself shall be in "raw" uncompressed 8-bit grayscale format. No format is prescribed for the enrollment data.

The fields **latent_finger->width** and **latent_finger->height** specify the width and height of the latent image in pixels. The field **latent_finger->resolution** specifies the horizontal and vertical resolution of the latent image in pixels per centimeter (e.g. 500 pixels per inch is specified as 197 ppcm ; 1000 ppi is specified as 394 ppcm). The fields **latent_finger->impression_type** and **latent_finger->finger_position** will always be set equal to 0.

The function may be optionally supplied with a "region of interest" in the form of an image mask. In cases where no "region of interest" information is provided, the input **roi_mask** parameter shall be a *NULL* pointer. Otherwise, **roi_mask** shall point to a "raw" uncompressed raw 8-bit grayscale image with the same dimensions as the latent fingerprint image. The region (or regions) of interest in the latent fingerprint image are identified by the corresponding *x,y* locations in the **roi_mask** having non-zero pixels.

The **orientation** parameter specifies the estimated angle of the fingerprint in degrees (0 to 359). The **offset** (0 to 180) specifies the offset (+ or -) in degrees around this angle of allowable variance. Taken together these values inform the SDK as to fingerprint image's range of rotational variance which may be useful to the matching algorithm. The angle is expressed in standard mathematical format, with zero degrees to the right and angles increasing in the counterclockwise direction. Thus "upright" fingerprint images are said to have an orientation of 90 degrees. For example, if **orientation** and **offset** are specified as 75 and 5 respectively, the fingerprint image is estimated to have an orientation between 70 and 80 degrees. The **offset** 180 will only be used in conjunction with an **orientation** of 90 to convey complete uncertainty as to the fingerprint's orientation, and in that case full rotational variance (0 to 359) shall be assumed.

The memory for **enrolled_latent** is allocated prior to the call (i.e. by the application linked with the SDK) as a pre-zeroized 10 megabyte array.

Upon return from this function, **enroll_length** shall be set by the SDK to the length (in bytes) of the enrollment data stored in **enrolled_latent**. The memory for **enroll_length** is allocated by the caller prior to calling this function.

Failure-to-enroll a latent shall result in a non-zero return code and upon return from this function the enrollment data written to **enrolled_latent** shall contain non-zero length data defined by the SDK as representing "null enrollment data." This "null enrollment data" shall be usable in subsequent searches for the corresponding latent, and result in the output of a candidate list with all entries set to 0.

Note that during the call to this function the directory containing the current background and its contents are read-only.

Parameters

    latent_finger **(input)**: Pointer to a latent fingerprint image record.

    roi_mask **(input)**: Pointer to optional image mask identifying ROI(s).

    orientation **(input)**: The estimated orientation (in degrees) of the latent fingerprint.

    offset **(input)**: The range of variance (in degrees) + or - the orientation.

    enrolled_latent **(output)**: Pointer to memory block receiving the enrollment data.

    enroll_length **(output)**: Pointer to length of enrolled_latent in bytes.

Return Value

This function returns zero on success or a documented non-zero error code on failure.

C.2.3.4 Image Search

```
INT32
image_search(const BYTE        *enrolled_latent,
             CANDIDATE_LIST    *candidates,
             char              *error_msg);
```

Description

This function searches the current background (as selected by set_background()) for zero or more candidates matching the input enrolled_latent parameter. The selection of features on which to match is entirely at the discretion of the SDK. Note that during the call to this function the directory containing the current background and its contents are read-only.

When this function is called, the candidates parameter will point to a pre-initialized CANDIDATE_LIST (see 2.1 above) with candidates->num_entries set equal to M, the number of background records (N) multiplied by 10 (i.e. M = N x 10), and candidates->list pointing to a pre-allocated M length array of (pre-zeroized) CANDIDATE structures.

During execution of this function the SDK shall modify the CANDIDATE_LIST structure such that candidates->num_entries is set equal to the number of candidates found (S), and the first S members of the array specified by candidates->list contain all candidate information. In other words, the first S structures of type CANDIDATE (see 2.1 above) pointed to by candidates->list shall contain the original background record file index, finger position, similarity score, and probability (range 0 to 100) for each candidate found by the search. For Phase I and II testing, the number of candidates found, S, shall equal 50 (and M will always be greater than 50). Additionally, the CANDIDATE structures in candidates->list shall be stored in decreasing order of similarity_score Note that before returning from this function the SDK ~~must~~ set candidates->num_entries equal to 50, even if less than 50 candidates are actually written to candidates->list . In the event that less than 50 candidates are actually written to candidates->list , the pre-zeroized CANDIDATE structures in the array will effectively provide "padding" (with NULL candidates) to the required length of 50.

The background_indexfield for each CANDIDATE shall be set equal to the relative offset of the original ten-print record file processed by enroll_background(). The finger_positionfor each CANDIDATE shall be set equal to the finger position information extracted from its associated ten-print record file. And the similarity_scorefor each CANDIDATE shall be set to a value greater than or equal to 0 which represents the similarity of the input latent finger image to the respective candidate finger image in the background. Note that any background fingerprint images not represented by an entry in candidates->list shall be implicitly assigned a similarity score equal to 0.

The probability field for each CANDIDATE shall be set equal to the probability (0-100) that the candidate is a "likely hit."

Non-fatal error conditions shall be tolerated and shall ~~not~~ result in pre-mature halting (i.e. non-completion of the search). These error conditions include encountering "gaps" in the background resulting from prior failure-to-enroll (FTE) events, and searching with an enrolled_latent containing "null enrollment data." In the latter case, the candidate list returned shall have all entries set to 0. If any of the above non-fatal error conditions are encountered, the SDK may optionally return a documented non-zero warning code (after completing the search), though this is not required.

Duplicate CANDIDATE entries or entries whose background_indexfield values are out of range (i.e. not between 1 and the N inclusive) shall not be accepted.

Upon entry the error_msg parameter will point to a pre-allocated and pre-zeroized string buffer of length 513 bytes (512 + 1 for the NULL terminator) that the SDK may use for outputting detailed information regarding fatal errors which have occurred (signaled by a non-zero return code). This may be useful for debugging any problems that might occur after the SDK is received by NIST.

Optionally, the quality of the latent print, the number of minutiae found in the latent print, the number of latent minutiae matching each candidate print, and the quality of the each candidate print may be returned (respectively) via the fields candidates->latent_quality , candidates->num_latent_minutiae , candidate->num_matching_minutiae , and candidate-

>candidate_quality . If image quality values are supplied for either the latent or candidate print, the table below indicates the required range of values and their associated meanings:

Image Quality Value	Description
20	Poor
40	Fair
60	Good
80	Very Good
100	Excellent

Note 1: Matcher architectures in which "advanced matchers" are selectively invoked (depending upon initial screening results for the latent) are allowed. The SDK might decide to invoke (call) computationally intensive matchers only for those comparisons which show initial good results. However, the SDK must decide if the additional features (if any) used by these "advanced matchers" will be written to persistent storage during the call to enroll_background().

Note 2: Since it may not be possible to keep all background images in memory, it might be necessary for the software to repeatedly retrieve the data from disk, and this extra fetch time will be included in the execution time measurement.

Note 3: The candidate list shall only depend on the inputs to this function and the currently selected background (not on any previous results from this function). Thus, identical inputs and background shall produce identical candidate lists independent of all prior calls to this function.

Parameters

enrolled_latent (**input**): Pointer to the latent image's enrollment data.

candidates (**input/output**): A list of candidates matching the latent fingerprint image.

error_msg (**output**): Pointer to a detailed error message string.

Return Value

This function returns zero on success or a documented non-zero error code on failure.

C.2.4 Error Codes and Handling

The participant shall provide documentation of all (non-zero) error or warning return codes (see section C.3.3, Documentation).

The application should include error/exception handling so that in the case of a fatal error, the return code is still provided to the calling application.

All messages which convey errors, warnings or other information shall be suppressed. Information supplemental to the documented error codes returned by the SDK shall be conveyed via the error_msg parameter (see 2.3 above) only.

At minimum the following return codes shall be used.

Return code	Function	Explanation
0	All	Success
-1	extract_image_data()	unable to open file

-2	extract_image_data()	Incorrect file format
-3	extract_image_data()	error parsing ten-print file
-4	extract_image_data()	error decompressing image
-5	extract_image_data()	insufficient memory error
-6	extract_image_data()	unspecified error
100	enroll_background()	enrollment directory not found
101	enroll_background()	error extracting image(s) from ten print
102	enroll_background()	error writing enrollment data
103	enroll_background()	insufficient memory error
200	set_background()	enrollment directory not found
300	enroll_latent()	image size not supported
301	enroll_latent()	image resolution not supported
302	enroll_latent()	insufficient features found in latent
400	image_search()	enrollment directory not set
401	image_search()	insufficient memory available for search
402	image_search()	unable to access original ten-print record

C.3 Software and Documentation

C.3.1 SDK Library and Platform Requirements

Individual SDKs shall not include multiple "modes" of operation, or algorithm variations which require explicit activation by the calling application. If participants wish to separately compare the performance of such features, they must submit separate SDKs. Note that this requirement does not preclude implementation of internally (i.e. autonomously) selected modes or algorithm variations within a single SDK. Only such features requiring external selection by the calling application are forbidden.

Participants shall provide NIST with binary code only (i.e. no source code) – supporting files such as header (".h") files notwithstanding. It is preferred that the SDK be submitted in the form of a single static library file (i.e. ".LIB" for Windows or ".a" for Linux). However, dynamic/shared library files are permitted.

If dynamic/shared library files are submitted, it is preferred that the API interface specified by this document be implemented in a single "core" library file with the base filename 'liblatent' (for example, 'liblatent.dll' for Windows or 'liblatent.so' for Linux). Additional dynamic/shared library files may be submitted that support this "core" library file (i.e. the "core" library file may have dependencies implemented in these other libraries).

Note that dependencies on external dynamic/shared libraries such as compiler-specific development environment libraries are discouraged. If absolutely necessary, external libraries must be provided to NIST upon prior approval by the Test Liaison.

The SDK will be tested in non-interactive "batch" mode (i.e. without terminal support). Thus, the library code provided shall not use any interactive functions such as graphical user interface (GUI) calls, or any other calls which require terminal interaction (e.g. calls to "standard input" or "standard output").

The use of multi-threading by the SDK is encouraged as the NIST test platform includes dual-processor support. The SDK need not be "thread safe" as the NIST test driver itself is single threaded. If multi-threading is utilized by the SDK is shall be documented.

NIST will link the provided library file(s) to a C language test driver application (developed by NIST) using the GCC compiler (for Windows platforms Cygwin/GCC version 3.3.1 will be used; for Linux platforms GCC version 2.96 and GNU ld 2.11.90.0.8 will be used. All GCC compilers use Libc 6). For example,

```
gcc -o latenttest latenttest.c -L. -llatent
```

Participants are required to provide their library in a format that is linkable using GCC with the NIST test driver, which is compiled with GCC. All compilation and testing will be performed on x86 platforms running either Windows 2000 Professional SP4 (or higher) or Linux (kernel 2.4.7-10 or higher) dependent upon the operating system requirements of the SDK. Thus, participants are strongly advised to verify library-level compatibility with GCC (on an equivalent platform) prior to submitting their software to NIST to avoid linkage problems later on (e.g. symbol name and calling convention mismatches, incorrect binary file formats, etc.).

C.3.2 Installation and Usage

The SDK must install easily (i.e. one installation step with no participant interaction required) to be tested, and shall be executable on any number of machines without requiring additional machine-specific license control procedures or activation.

The SDK's usage shall be unlimited. No usage controls or limits based on licenses, execution date/time, number of executions, etc. shall be enforced by the SDK.

It is requested that the SDK be installable using simple file copy methods, and not require the use of a separate installation program. Contact the Test Liaison for prior approval if an installation program is absolutely necessary.

C.3.3 Documentation

Complete documentation of the SDK shall be provided, and shall detail any additional functionality or behavior beyond what is specified in this document. The documentation must define all error and warning codes.

Multi-threading behavior by an SDK shall be documented.

C.3.4 Speed Requirement

All times given assume the use of a 2.8GHz Pentium IV equivalent or faster processor. Time will be measured as "wall clock" elapsed time.

The average time to enroll a single background ten-print record shall take no more than 150 seconds (15 sec/image).

The average time to enroll a single latent image shall take no more than 600 seconds.

The average time to search a single background ten-print record shall take no more than 0.25 seconds (0.025 sec/image).

Appendix D - Complete Set of Accuracy Characteristics

This appendix presents an exhaustive catalog of the DET characteristics. This consists of sixteen computed from the SDKs' raw matcher score, and the remainder from the reported probability estimate. All providers' DET curves are displayed together. There is a separate set of four plots (for ranks 1, 10, 20, and 50) for each of the four tests.

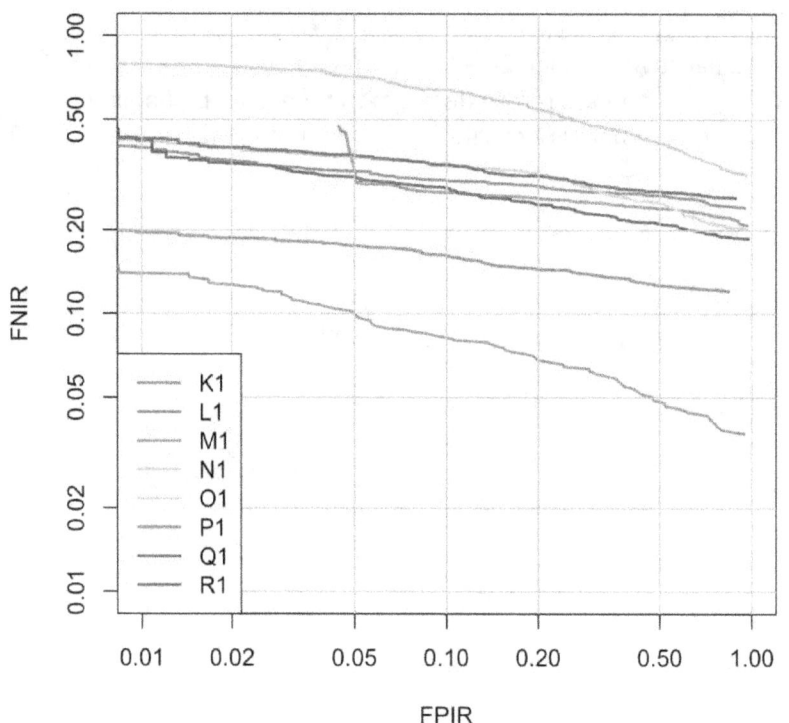

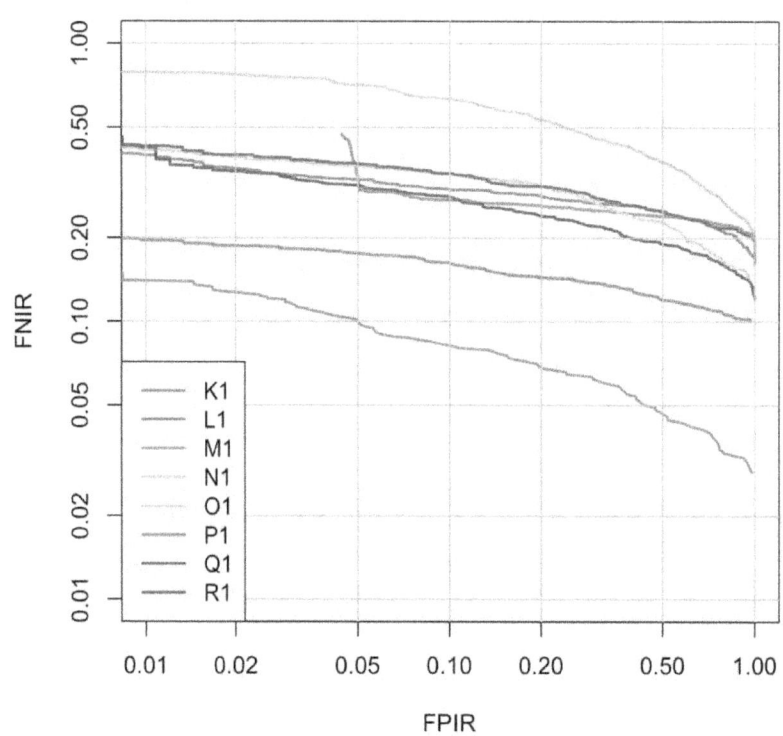

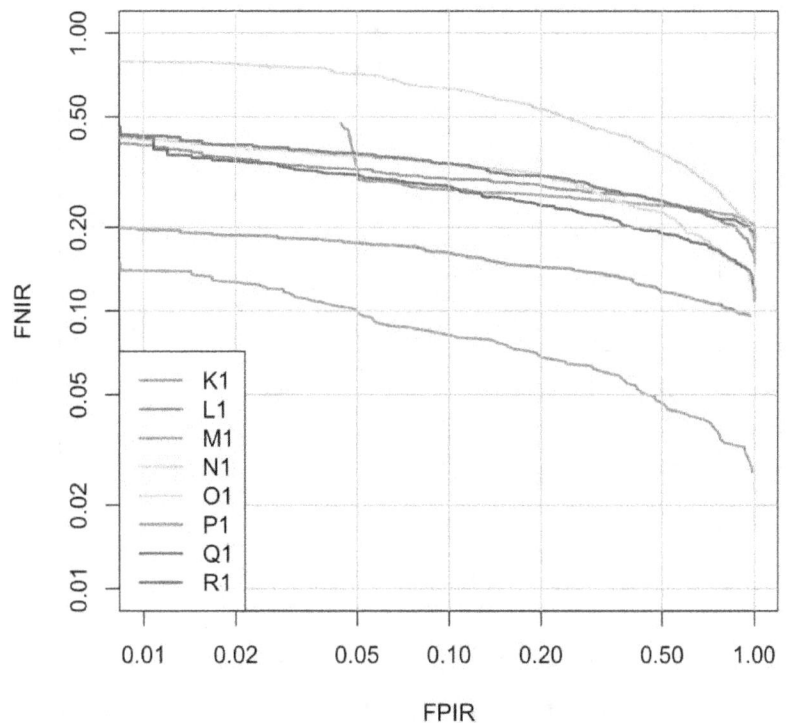

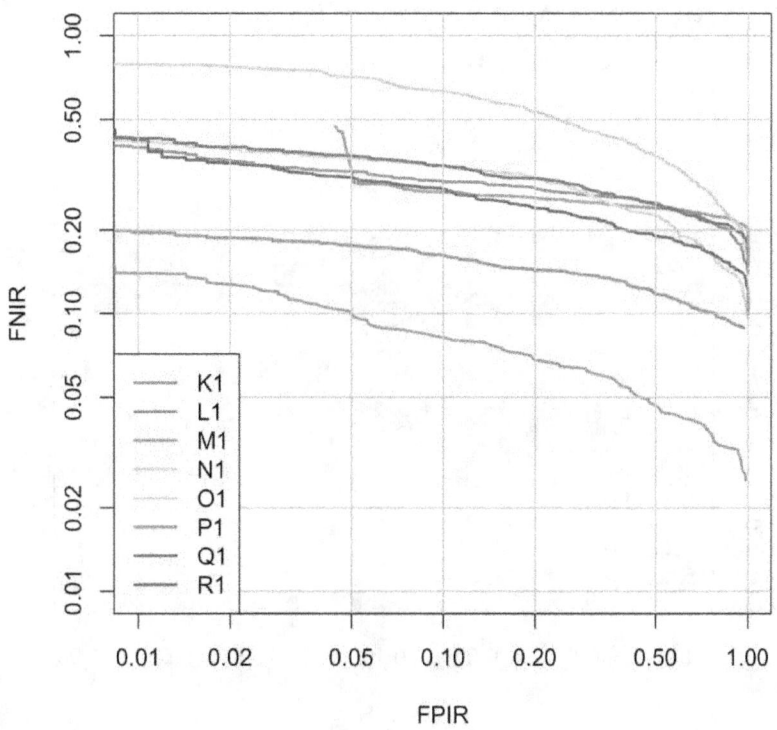

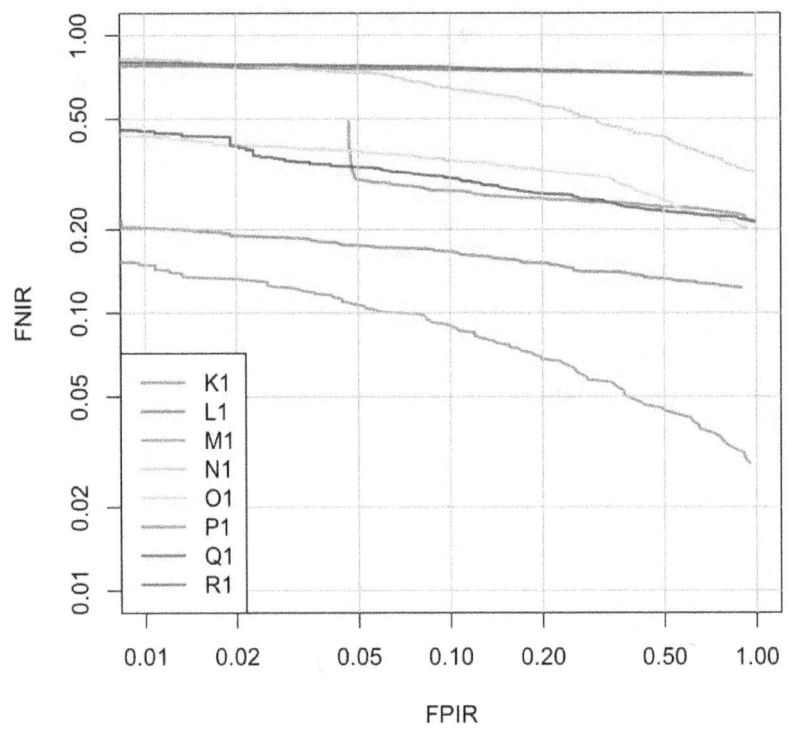
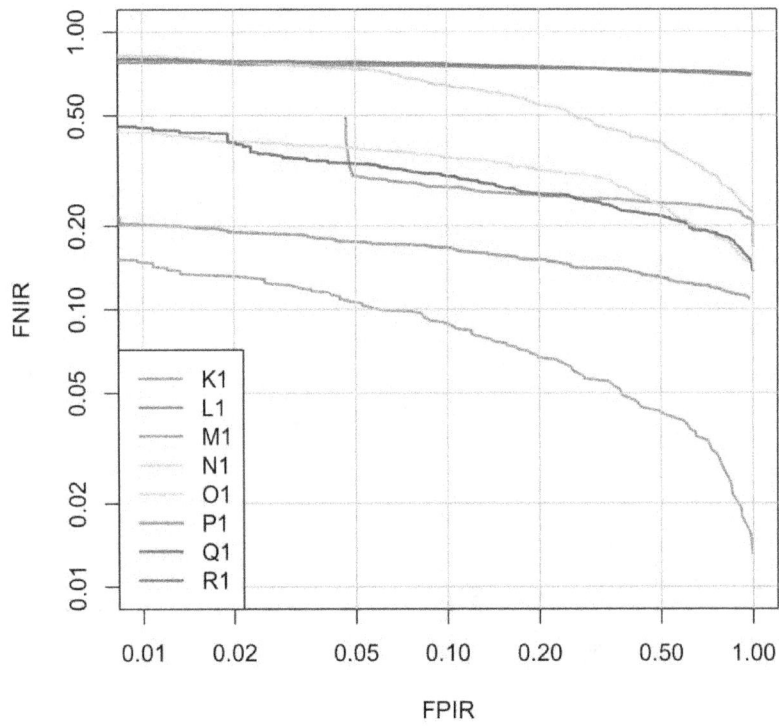

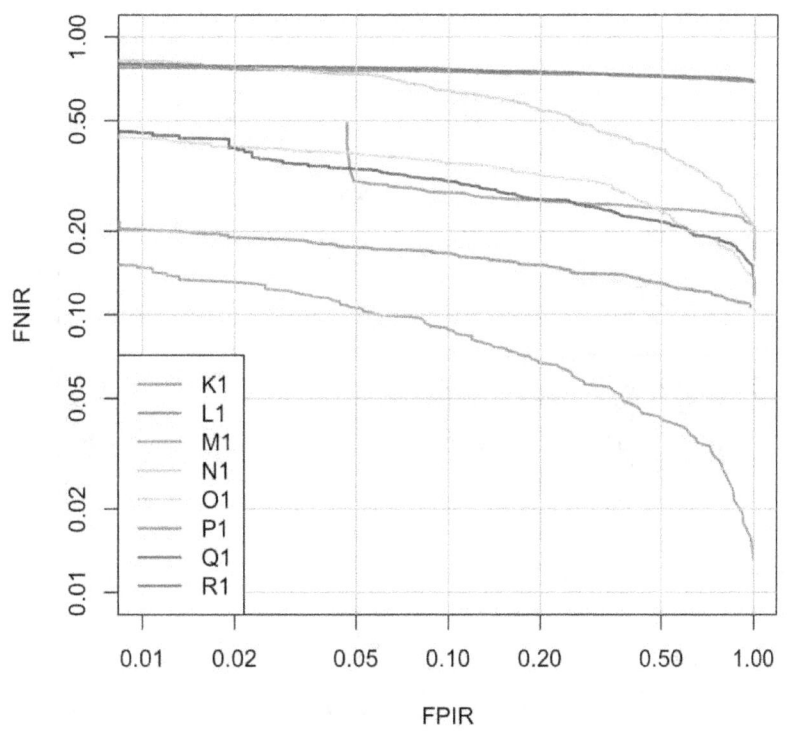

DET - rank 20 - resolution 1000ppi/gallery_10k

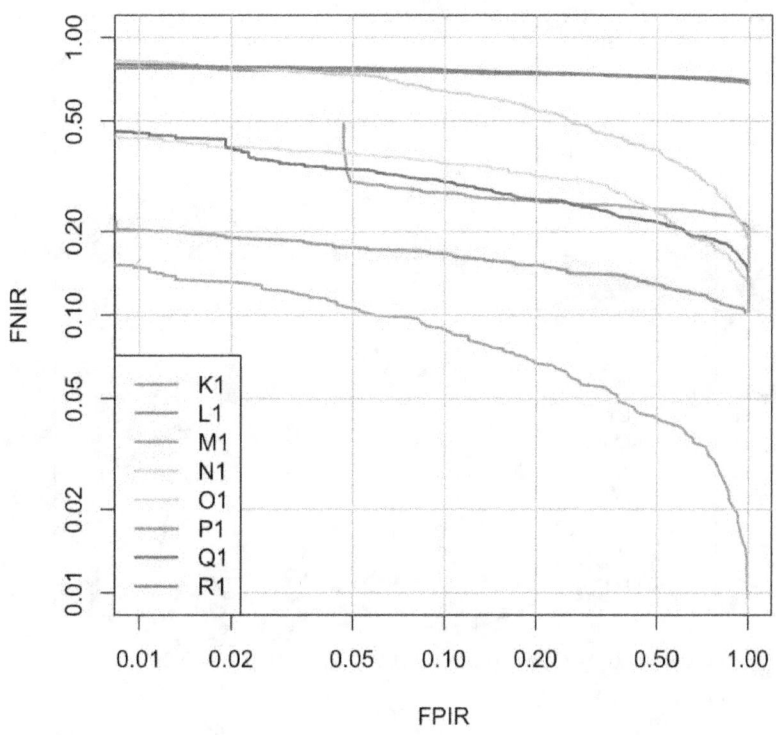

DET - rank 50 - resolution 1000ppi/gallery_10k

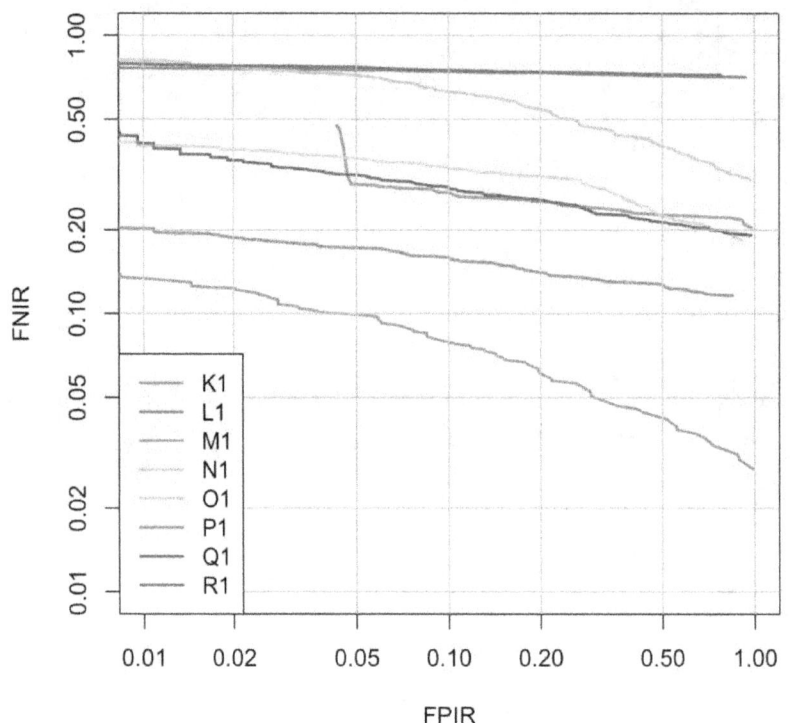

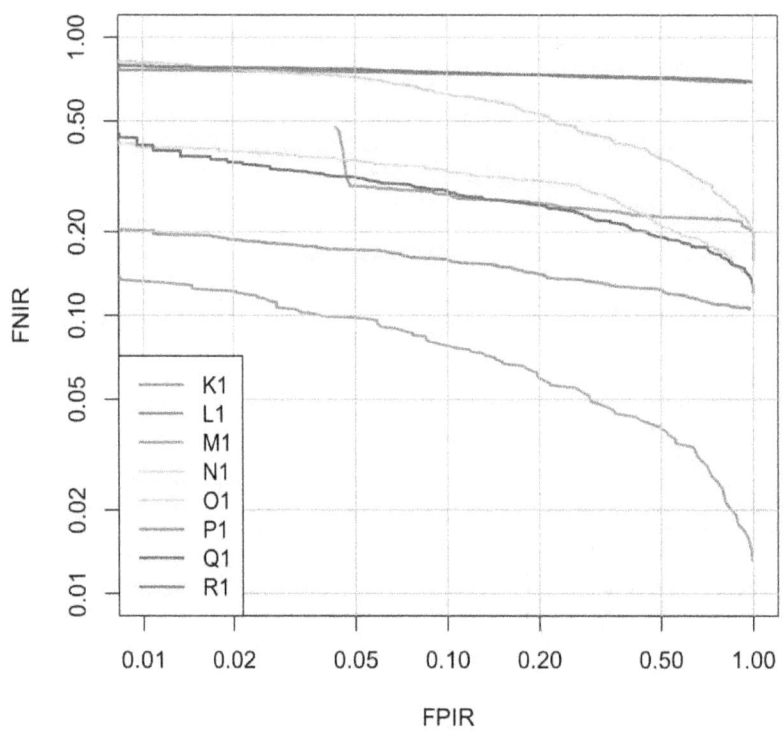

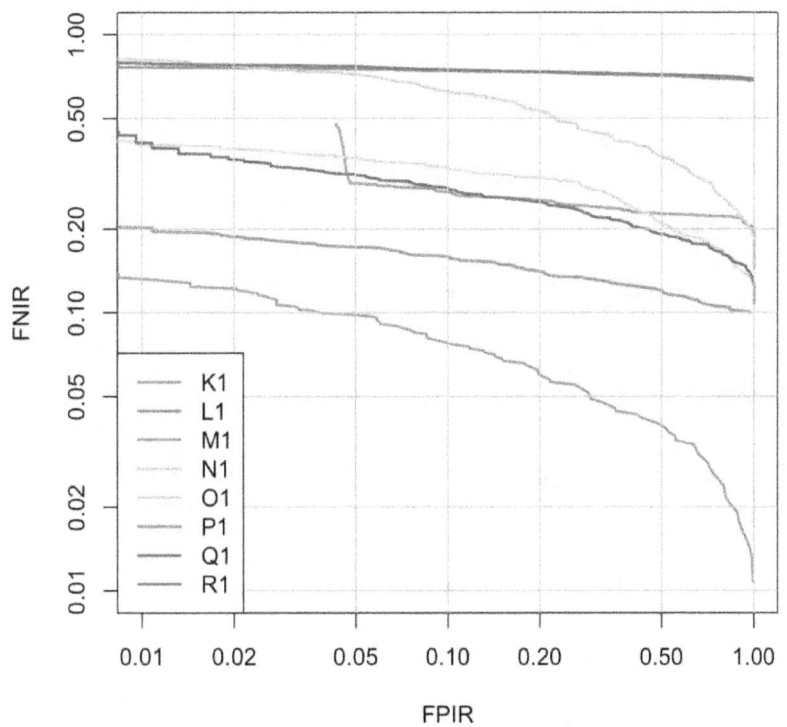

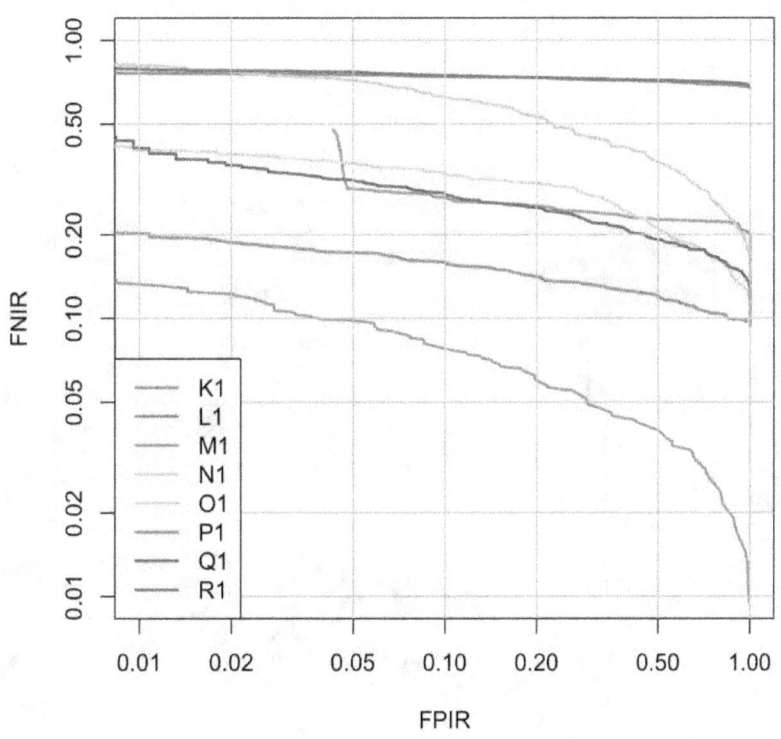

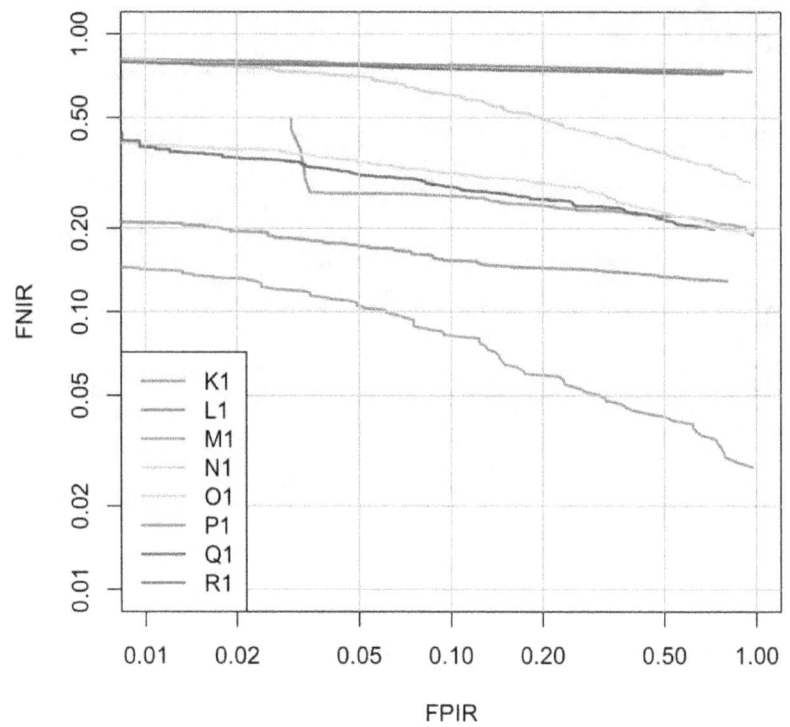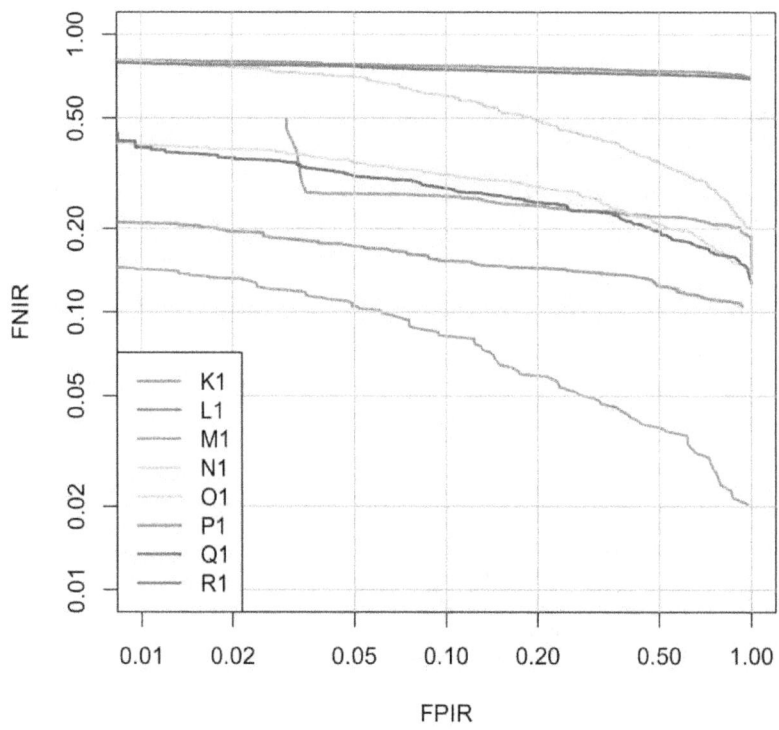

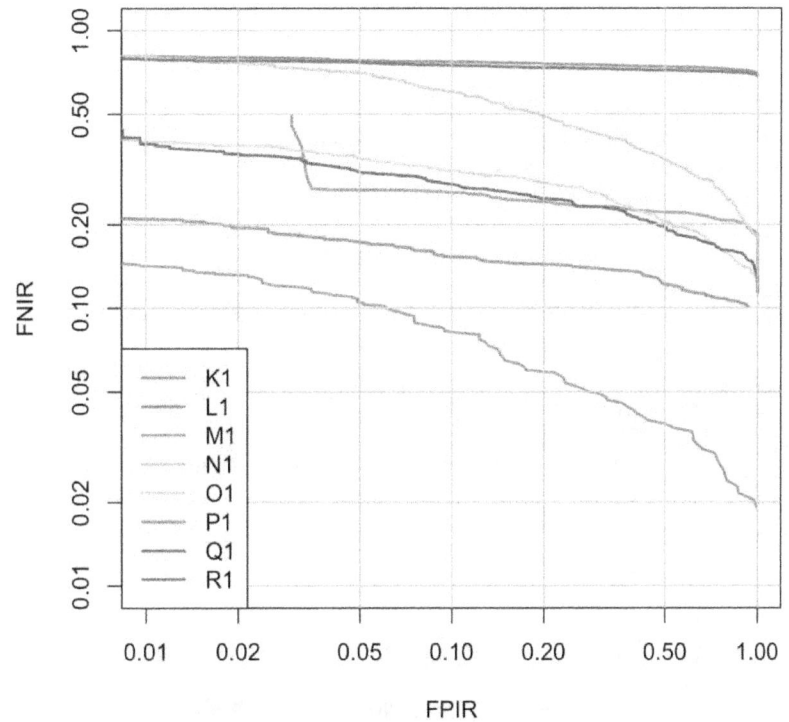
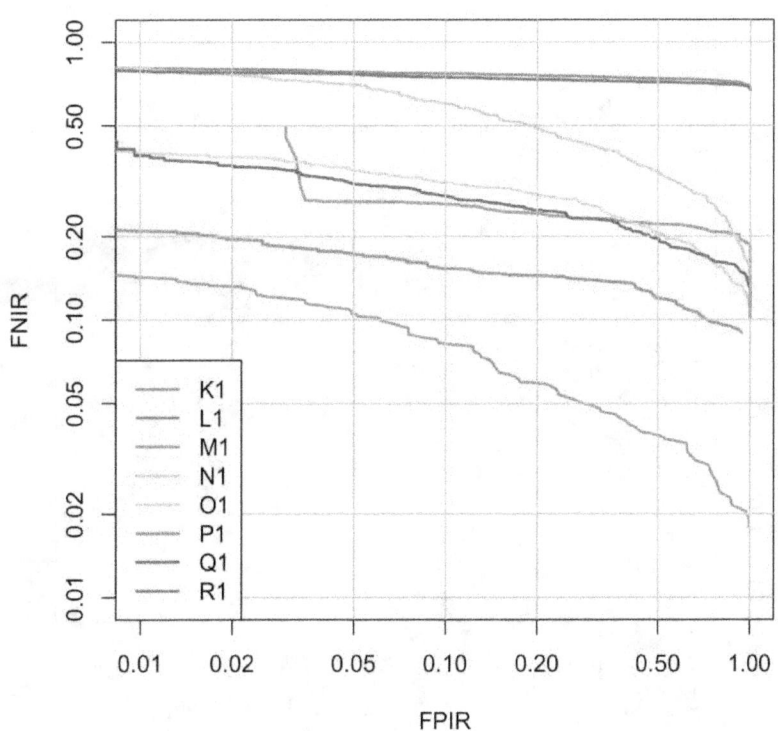

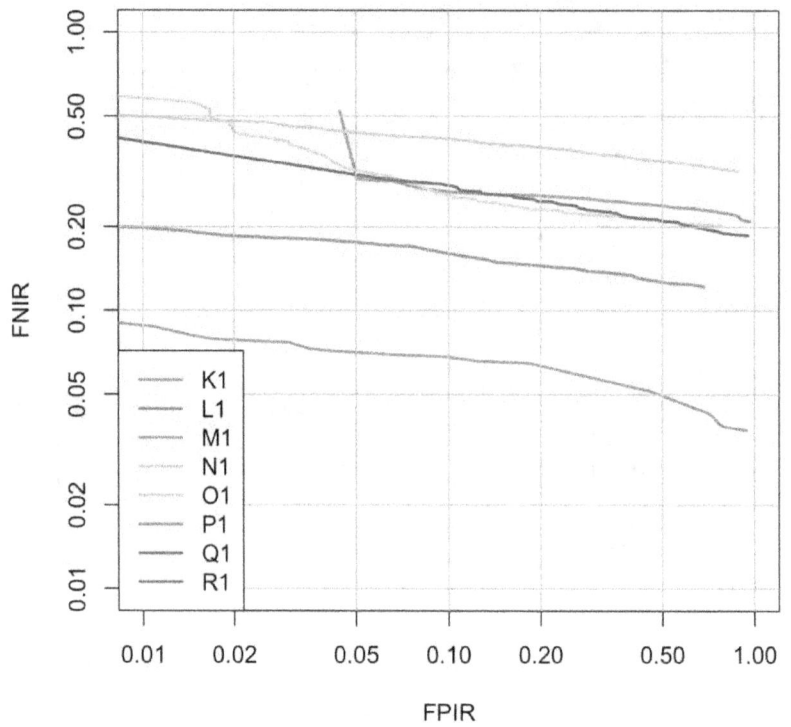
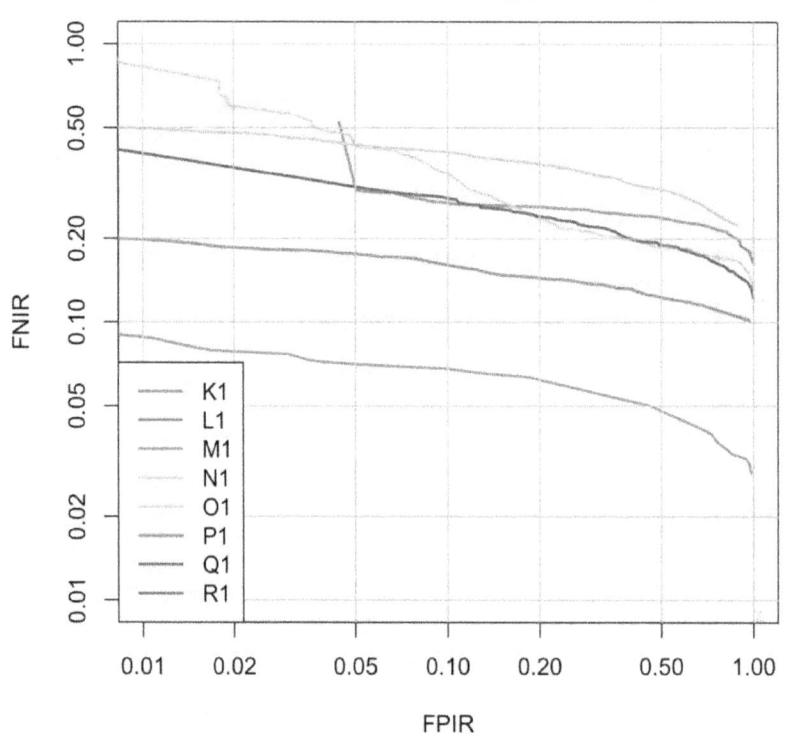

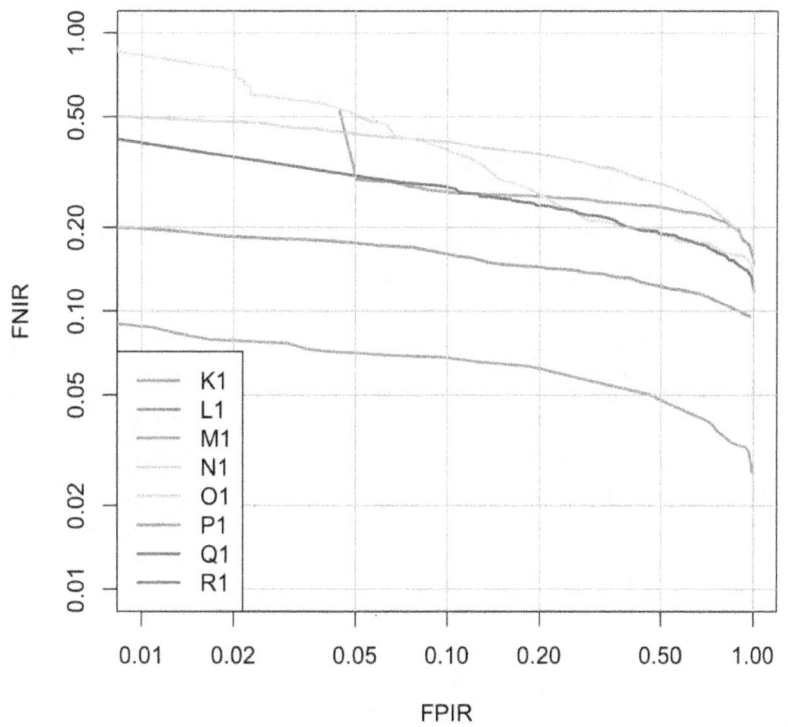
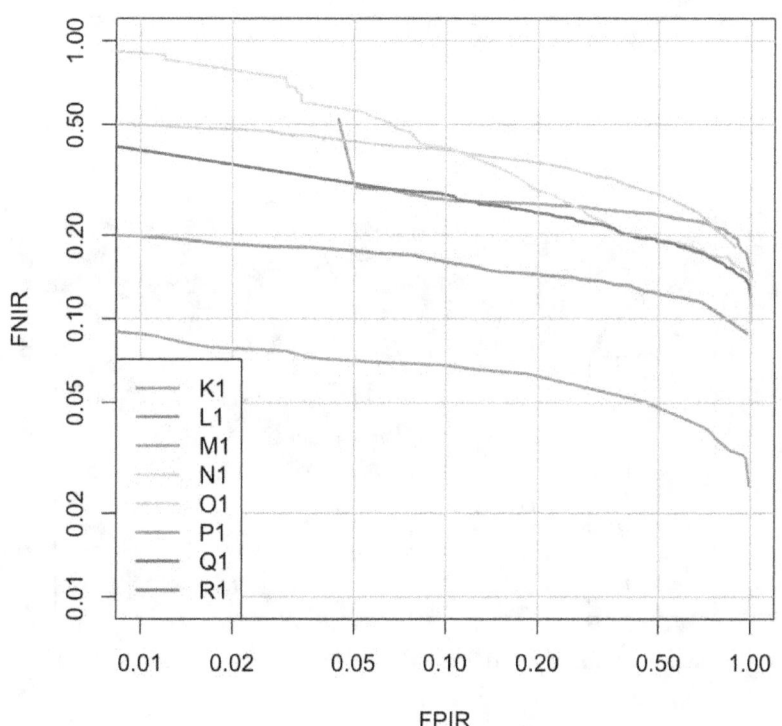

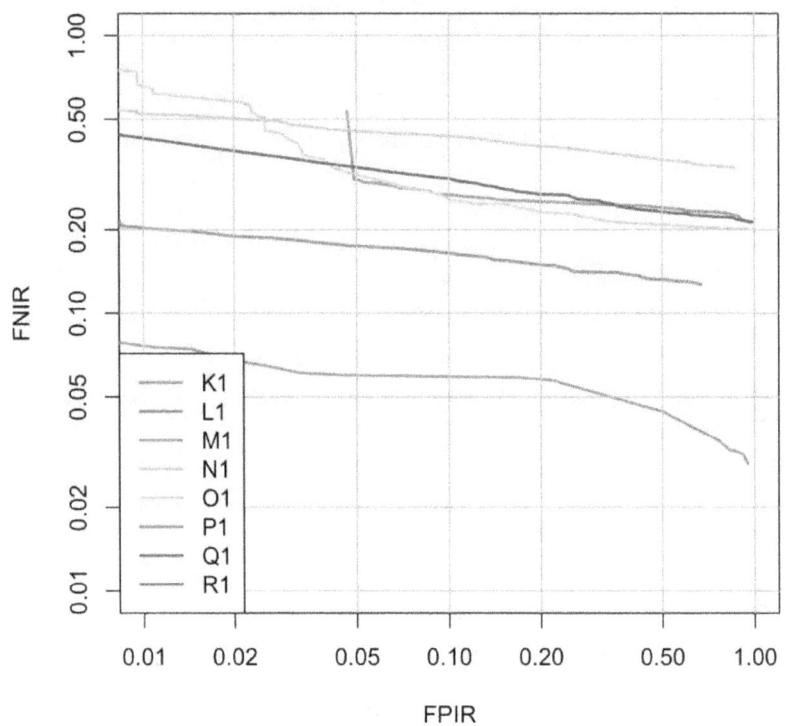

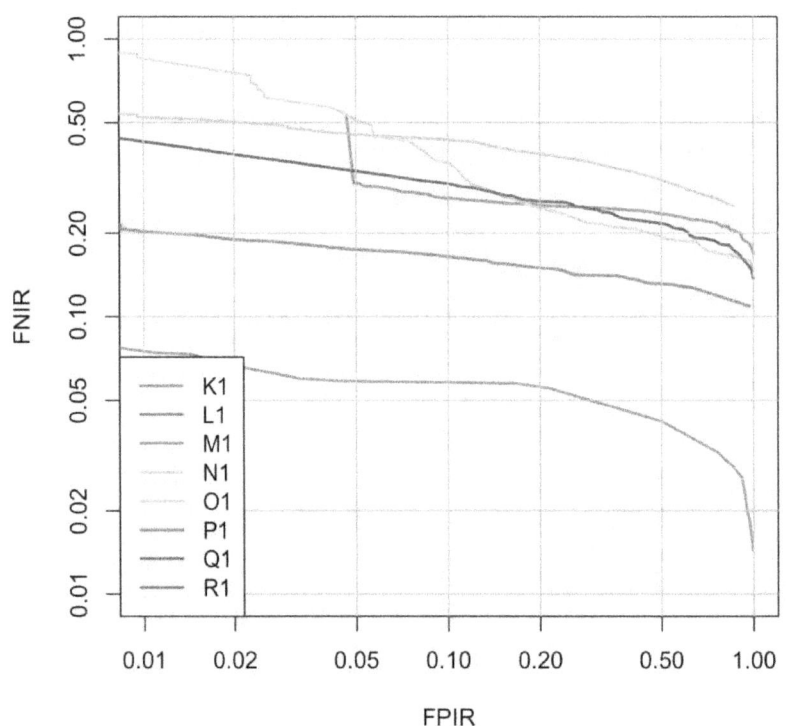

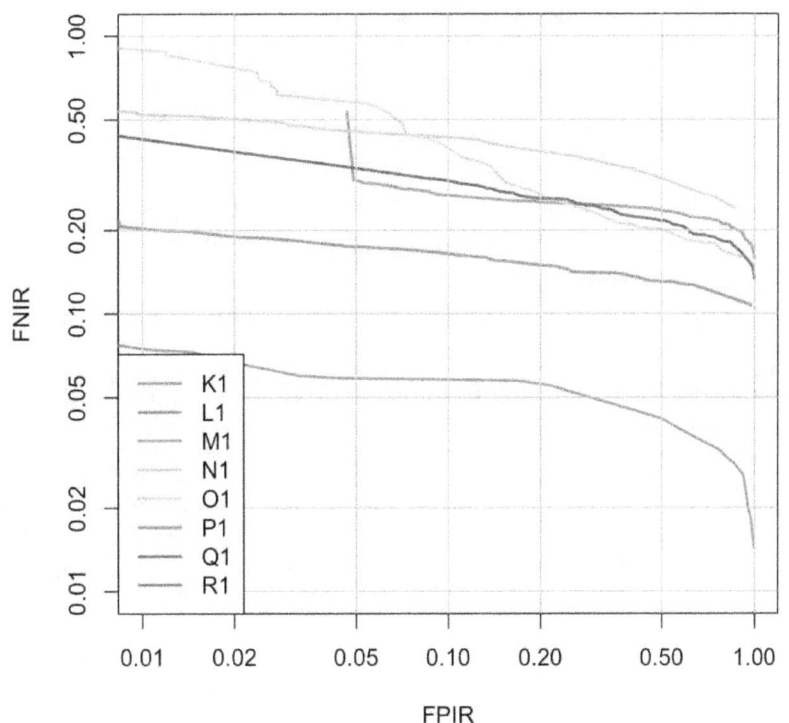

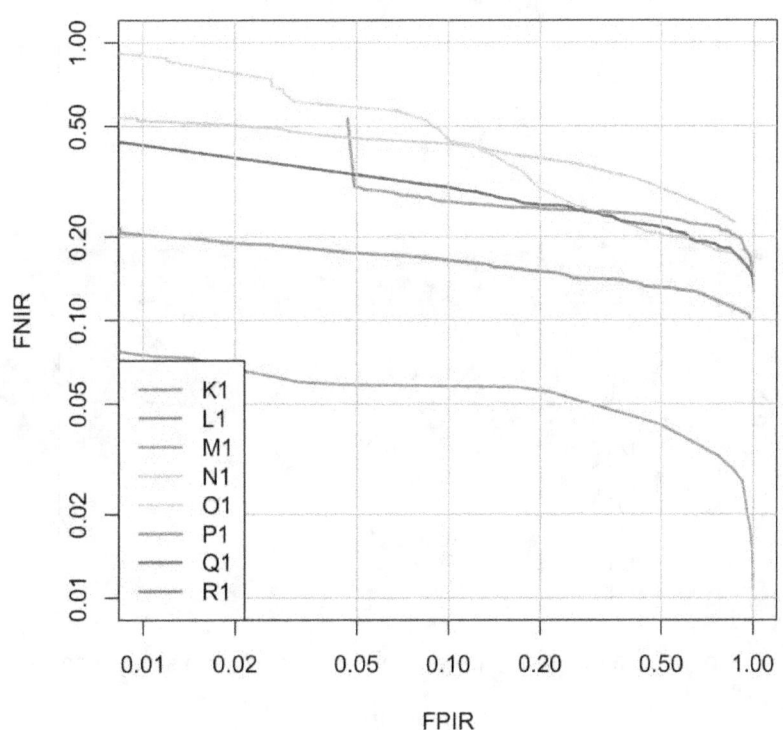

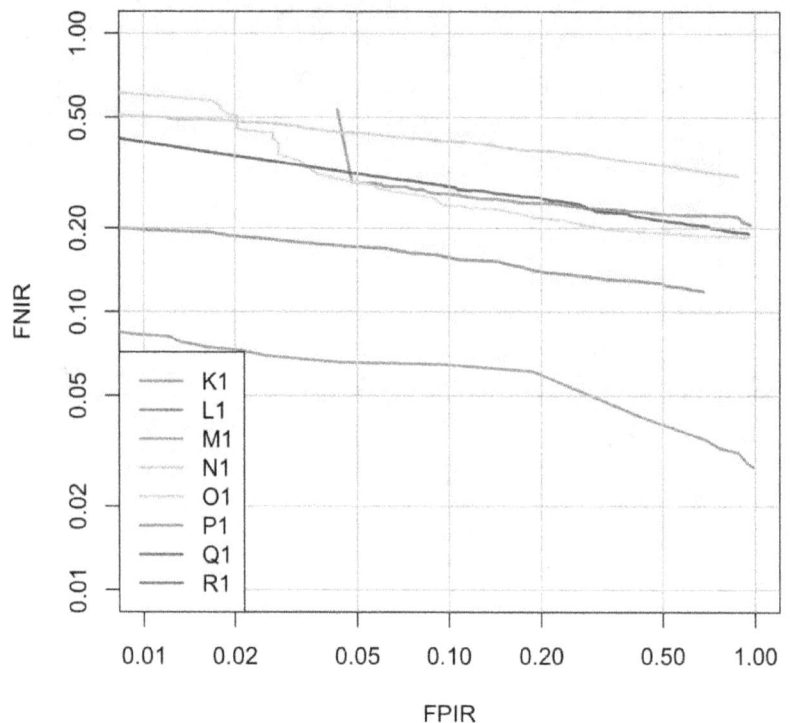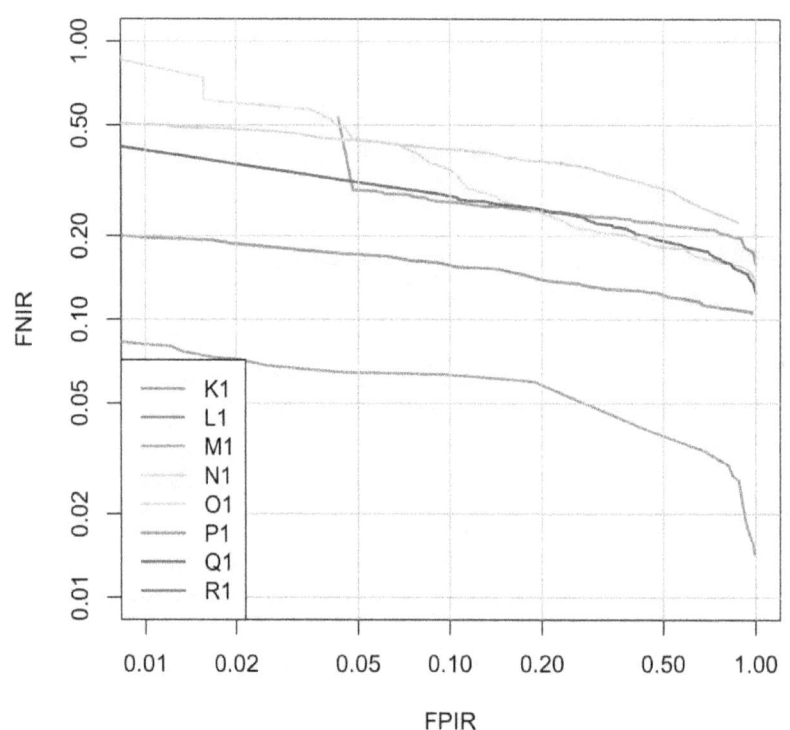

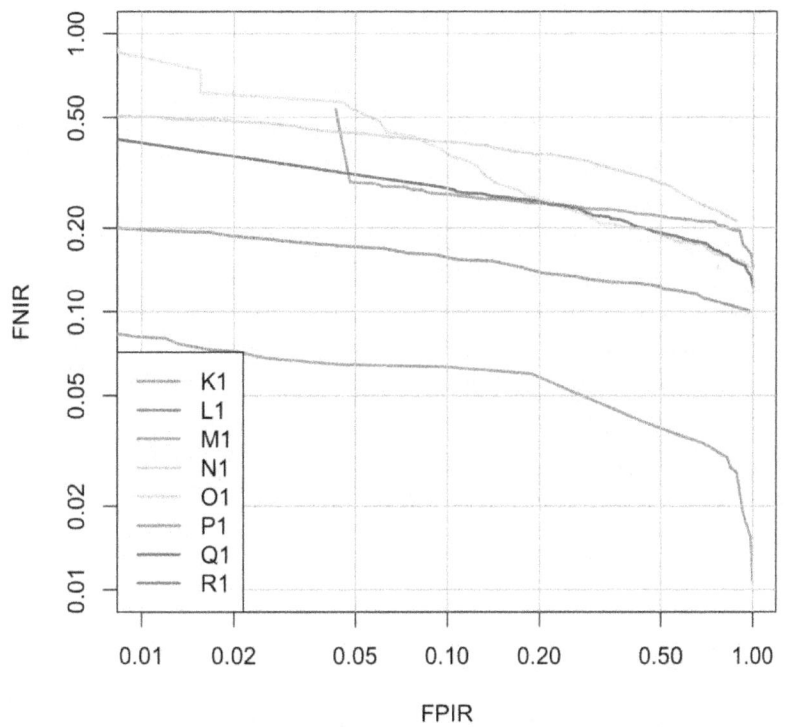

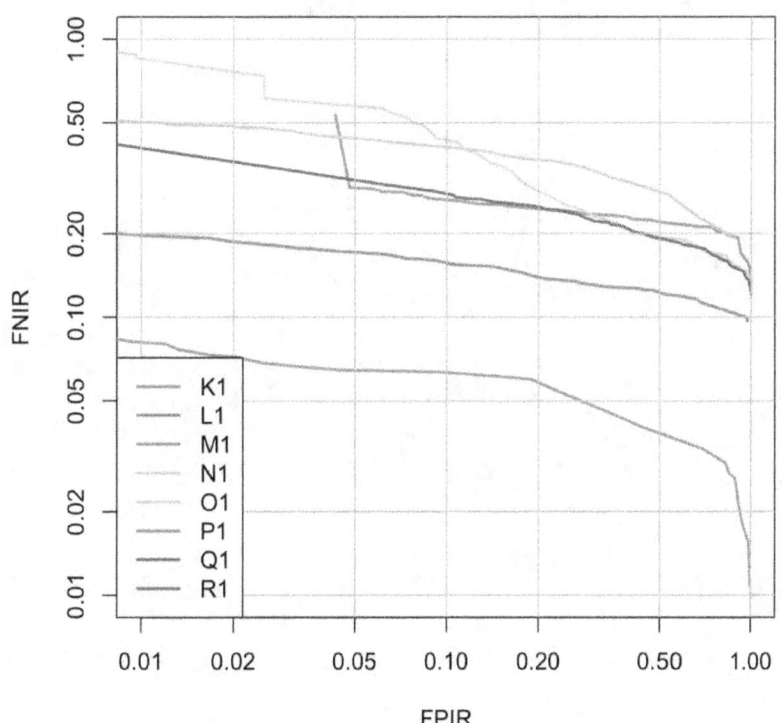

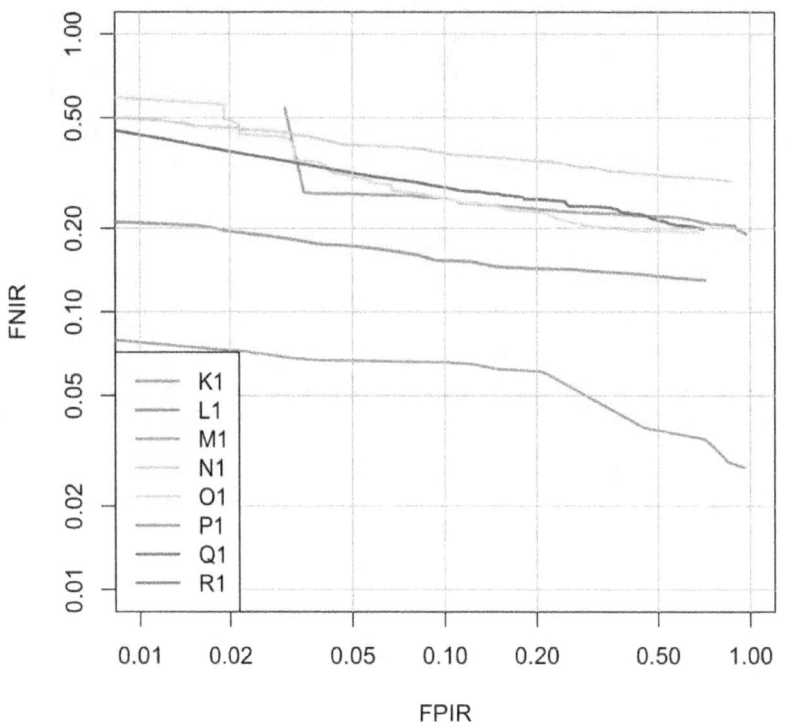

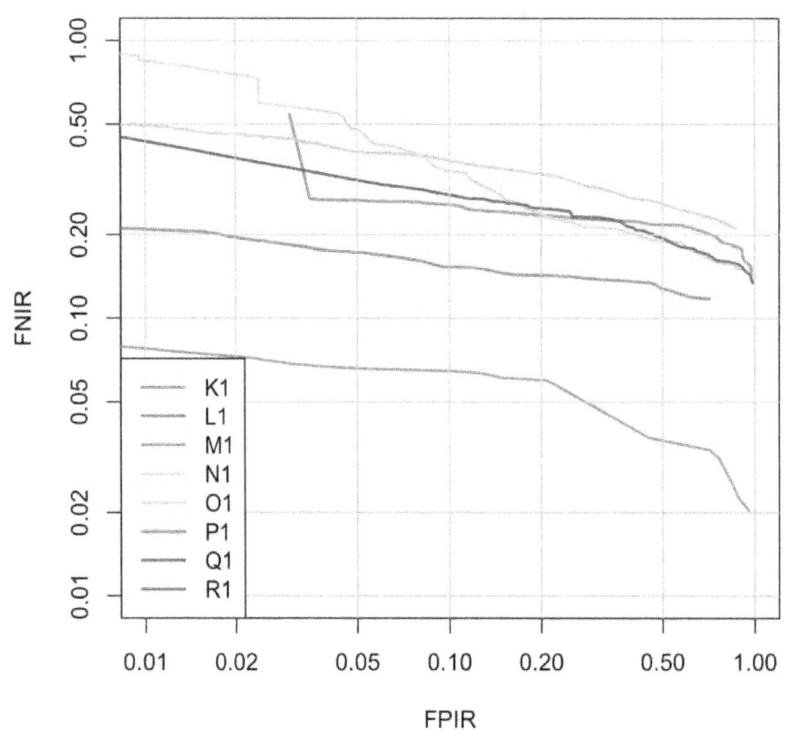

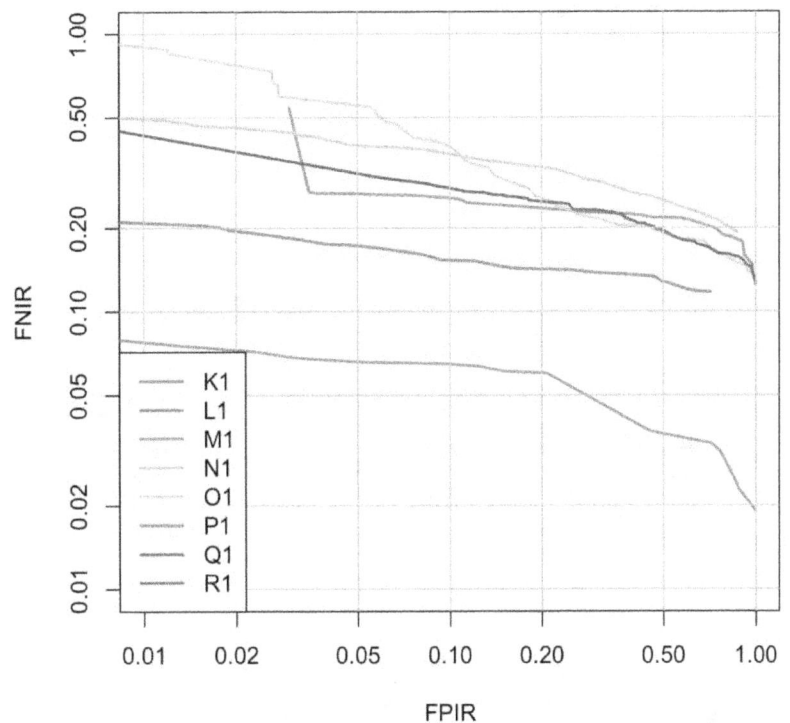

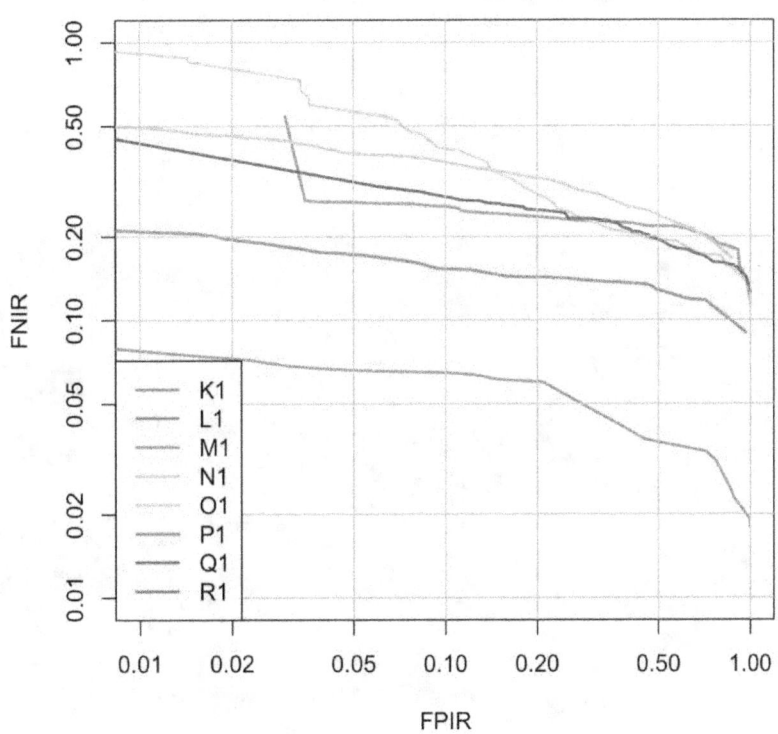